I0044538

Fundamentals of Hydrology

Fundamentals of Hydrology

Harvey Rogers

SYRAWOOD
PUBLISHING HOUSE
New York

Published by Syrawood Publishing House,
750 Third Avenue, 9th Floor,
New York, NY 10017, USA
www.syrawoodpublishinghouse.com

Fundamentals of Hydrology
Harvey Rogers

© 2020 Syrawood Publishing House

International Standard Book Number: 978-1-64740-025-5 (Hardback)

This book contains information obtained from authentic and highly regarded sources. All chapters are published with permission under the Creative Commons Attribution Share Alike License or equivalent. A wide variety of references are listed. Permissions and sources are indicated; for detailed attributions, please refer to the permissions page. Reasonable efforts have been made to publish reliable data and information, but the authors, editors and publisher cannot assume any responsibility for the validity of all materials or the consequences of their use.

Trademark Notice: Registered trademark of products or corporate names are used only for explanation and identification without intent to infringe.

Cataloging-in-Publication Data

Fundamentals of hydrology / Harvey Rogers.
    p. cm.
Includes bibliographical references and index.
ISBN 978-1-64740-025-5
1. Hydrology. 2. Hydrography. 3. Water. 4. Earth sciences. I. Rogers, Harvey.
GB661.2 .F86 2020
551.48--dc23

TABLE OF CONTENTS

PREFACE

The branch of science which studies the distribution of water as well as its quality and movement is known as hydrology. It is also involved in the collection and analysis of data for solving water related problems like natural disasters and wastewater treatment. Some of the different areas studied within this field are water cycle, environmental watershed sustainability and water resources. There are various sub-disciplines within hydrology such as groundwater hydrology, marine hydrology, chemical hydrology, isotope hydrology and hydrogeology. This discipline finds application in a variety of areas like in determining the water balance of a region, providing drinking water and designing bridges. The book aims to shed light on some of the unexplored aspects of hydrology. It presents this complex subject in the most comprehensible and easy to understand language. This textbook will serve as a reference to a broad spectrum of readers.

A short introduction to every chapter is written below to provide an overview of the content of the book:

Chapter 1 - The branch of science which studies the distribution, movement and quality of water on Earth as well as other planets is termed as hydrology. An important area of study within this field is the cryosphere, which refers to the water which exists in the frozen form on Earth. The topics elaborated in this chapter will help in gaining a better perspective about hydrology and its branches.; **Chapter 2** - The continuous movement of water under, over and on the surface of the Earth is known as the hydrological cycle or the water cycle. The flow of water into and out of a system is described using an equation called a water balance equation. This chapter closely examines the key concepts of water balance and the hydrological cycle to provide an extensive understanding of the subject.; **Chapter 3** - The branch of hydrology which seeks to study the distribution and movement of groundwater in the rocks and soil is referred to as groundwater hydrology. Surface water hydrology deals with the study of water above the surface of the Earth. This chapter has been carefully written to provide an easy understanding of the varied facets of groundwater and surface hydrology.; **Chapter 4** - The branch of science which seeks to study the interactions between the different ecological systems and water is termed as ecohydrology. Isotope hydrology refers to the branch of hydrology which is involved in estimating the origins and age of water through isotope dating. The topics elaborated in this chapter will help in gaining a better perspective about ecohydrology and isotope hydrology.; **Chapter 5** - The removal of sub-surface or water from an area through artificial or natural means is termed as drainage. A watershed, also known as a drainage basin, refers to a piece of land where rain or snow collects and flows off into a common outlet. This chapter has been carefully written to provide an easy understanding of the varied facets of drainage, drainage systems, watershed and watershed management.; **Chapter 6** - Wetland hydrology deals with the study of the flow of water into and out of a wetland along with its interaction with other site factors. The branch of study which seeks to analyze the distribution, storage and quality of water along with the hydrological processes in forest-dominated ecosystems is called forest hydrology. This chapter discusses in detail the hydrology of wetlands and forests as well as the different types of forests.

I extend my sincere thanks to the publisher for considering me worthy of this task. Finally, I thank my family for being a source of support and help.

Harvey Rogers

Chapter 1

Introduction to Hydrology

The branch of science which studies the distribution, movement and quality of water on Earth as well as other planets is termed as hydrology. An important area of study within this field is the cryosphere, which refers to the water which exists in the frozen form on Earth. The topics elaborated in this chapter will help in gaining a better perspective about hydrology and its branches.

Hydrology is the science that deals with all aspects of the water available on the earth. It includes study of occurrence of water, its properties, its distribution and circulation and also its effects on the living beings and their surroundings. It is not entirely a pure science because it has many practical applications and it utilizes knowledge of other sciences greatly.

Broadly, the whole subject matter can be expressed in the form of a mathematical equation.

The equation is:

$$P = R + L \text{ or}$$

Precipitation = Runoff + Losses.

In the above equation precipitation indicates total supply of water from all forms of falling moisture and mainly includes rainfall and snowfall. The runoff represents surplus water that flows over the surface to join some river or sea.

The term losses includes that portion of water which goes to atmosphere and underground by the processes like evaporation and percolation respectively. For practical reasons hydrology does not cover all studies- of oceans and medical uses of water.

After studying this equation with the background of hydrologic cycle it will be clear that the term losses never implies that this water is lost and cannot be used again. It is the water which temporarily disappears from view (e.g. evaporation, seepage, etc.) and given favourable conditions, reappears to perform various duties. Hence, it is necessary to study all the three terms of the equation, namely rainall, runoff, and losses.

Branches of Hydrology

Surface Hydrology

It is the study of bodies of water on or near the earth's surface. rivers, dams, lakes, and reservoirs are all part of this area of study, which further includes the systems used in recreational activity and transportation. Surface hydrology addresses issues pertaining to eroding soils and streams due to surface flow. Flooding, nutrient runoff, and pollutants are a few of the effects addressed, as well as the destruction of civil constructions such as dams. Methods of hydraulic and hydrologic design regulation are also undertaken in this field of study, as researchers simulate the long and short-term effects of anthropogenically manipulated surface water forms.

Isotope Hydrology

It is the study of the isotopic signatures of water. This subfield of hydrology utilizes isotopic dating to determine the origin and age of water throughout its movement within the hydrologic cycle. Isotopic dating involves measuring the levels of deviation in the isotopes of oxygen and hydrogen in water. Researchers are able to determine groundwater dated as far back as the Ice Age by using these techniques. Isotope hydrology deals with water usage policy, mapping aquifers, conservation of water resources, and maintaining pollution levels.

Hydromorphology

Hydromorphology is the study of the physical characteristics of bodies of water on the Earth's surface, including river basins, channels, streams, and lakes. Water quality, levels of pollution, and biological components needed for ecological system maintenance are a few areas assessed when classifying water systems. Hydromorphology studies the dynamics of groundwater flow into channels, lakes, and streams. It measures flow patterns and geometry as well as routing flows to avoid flooding or drought.

Hydrometeorology

Hydrometeorology is the study of the transfer of water and energy between land and water body surfaces and the lower atmosphere. Hydrometeorology incorporates meteorology to solve hydrological problems. These problems include forecasting flood or drought, or determining water resources and the safety of dams. Hydrometeorologists try to determine, through empirical data or theory, how the dynamics of water in the atmosphere affect the greatest levels of precipitation reaching the ground. The domain of hydrometeorology in the physical sciences is not very clearly defined, as it involves cloud physics, climatology, weather forecasting, and hydrology, to name a few.

Hydrogeology

Hydrogeology is the study of the distribution and movement of water in aquifers and shallow porous media—that is, the porous layers of rock, sand, silt, and gravel below the Earth's surface. Hydrogeology examines the rate of diffusion of water through these media as the water moves down its energy gradient. The flow of water in the shallow subsurface is also pertinent to the fields of soil science, agriculture, and civil engineering. The flow of water and other fluids (hydrocarbons and geothermal fluids) in deeper formations is relevant to the fields of geology, geophysics, and petroleum geology. Geohydrology is the area of geology that deals with the distribution and movement of groundwater in the soil and rocks of the Earth's crust. The term geohydrology is often used interchangeably.

Ecohydrology

Ecohydrology is the study of ecological processes in the hydrologic cycle. As these processes occur in the soil and plant foliage, ecohydrologists study how the hydrologic system affects plant physiology, soil moisture, and plant diversity and spatial orientation in various regions over a period of time. Ecohydrology has four main components: infiltration of precipitation into the soil, evapotranspiration, leakage of water into deeper portions of the soil not accessible to the plant, and runoff from the ground surface.

Forest Hydrology

Forest hydrology is an interdisciplinary applied science that intersects the disciplines of hydrology and forestry. It is primarily involved with the research and analysis of the effects of forest cover and changes of land use on water yield, quality and timing.

Hydrogeochemistry

Hydrogeochemistry is the science that uses the tools and principles of chemistry to explain the mechanisms behind major geological systems such as the earth's crust and its oceans.

Hydrography

Hydrography as it applies to the area of coastlines can be defined as ' (1) the description and study of seas, lakes, rivers and other waters. (2) The science of locating aids and dangers to navigation. (3) The description of physical properties of the waters of a region'. The term 'Hydrography' as it applies to the area of reclamation can be defined as ' Scientific study of physical aspects of all waters on the Earth's surface. Water features in 7.5-minute quads include lakes, shorelines, and drainage routing'.

Karst Topography

Karst is a geological formation shaped by the dissolution of a layer or layers of soluble bedrock usually carbonate rock such as limestone or dolomite, but also in gypsum. It has also been documented for weathering-resistant rocks, such as quartzite, given the right conditions. Karst topography is a geological formation shaped by the dissolution of a layer or layers of soluble bedrock, usually carbonate rock such as limestone or dolomite, but also in gypsum. It has also been documented for weathering-resistant rocks, such as quartzite, given the right conditions. Subterranean drainage may limit surface water with few to no rivers or lakes.

Many karst regions display distinctive surface features, with cenotes and sinkholes (also called dolines) being the most common. However, distinctive karst surface features may be completely absent where the soluble rock is mantled, such as by glacial debris or confined by one or more superimposed non-soluble rock strata. Some karst regions include thousands of caves, although evidence of caves large enough for human exploration is not a required characteristic of karst.

Hydrogeophysics

This branch has emerged over the last decade as one of the more challenging disciplines in near-surface geophysics, aiming to improve the simultaneous use of geophysical and hydrogeological measurements. It can be described as the use of geophysical measurements for mapping subsurface features, estimating properties and monitoring processes that are important to hydrological studies, such as those associated with water resources, seepage throughout the vadose zone, contaminant transport, and ecological/climate investigations.

Vadose Zone Hydrology

It is also termed the unsaturated zone is the part of Earth between the land surface and the top of the phreatic zone i.e. the position at which the groundwater (the water in the soil's pores) is at atmospheric pressure. Hence the vadose zone extends from the top of the ground surface to the water table. Water in the vadose zone has a pressure head less than atmospheric pressure, and is retained by a combination of adhesion (funiculary groundwater), and capillary action (capillary groundwater). If the vadose zone envelops soil, the water contained therein is termed soil moisture.

In fine grained soils, capillary action can cause the pores of the soil to be fully saturated above the water table at a pressure less than atmospheric. In such soils, therefore, the unsaturated zone is the upper section of the vadose zone and not identical to it. Movement of water within the vadose zone is studied within soil physics and hydrology, particularly hydrogeology, and is of importance to agriculture, contaminant transport, and flood control. The Richards equation is often used to mathematically describe the flow of water, which is based partially on Darcy's law. Groundwater recharge, which is an important process that refills aquifers, generally occurs through the vadose zone. from precipitation.

Coastal Zone Hydrology

There is no standard definition of what constitutes a 'coastal zone', but it could be said to be the interface between the land and water. These zones are dynamic areas with frequently changing biological, chemical and geological attributes and are of high economic significance, which is often subject to fast economic development, large population migrations and urban development.

Island Zone Hydrology

Island zone hydrology is a lens is a convex layer of fresh groundwater that floats on top of denser saltwater. It arises when rainwater seeps down through a soil surface and then gathers over a layer of seawater at or down to about five feet below sea level. Freshwater lenses are often found on small coral or limestone islands and atolls, where wells dug into them may be the only natural source of potable water.

Arid Zone Hydrology

Arid areas have distinctive hydrological features substantially different from those of humid areas. The term arid refers to severe lack of water in a region. The shortage of water prevents or hinders development and growth of animal and plant life. Arid areas lack vegetation and they are also called deserts. Arid zones are usually areas of scarce hydrological data. Any study requires an iterative approach to be adopted to develop water resources, with initial information often derived from other geographical regions which have similar aridity characteristics. Arid zones present special problems to hydrologists. Describing the hydrological balance is difficult as very little of the potential evaporation demand is satisfied. Only a small proportion of the rainfall becomes runoff except in extreme storms. The amount of rainfall tends to be highly variable in arid regions both in time and space. This leads to very high variability of runoff. Water, especially groundwater, is the key to life in semi-arid and arid zones.

Wadi Hydrology

Wadi is the Arabic word for ephemeral water courses in the arid regions, and they are a vital source of water in most arid and semi-arid countries. Catastrophic flash floods occurring in wadis are, on one hand, a threat to many communities and, on the other, major groundwater recharge sources after storms.

In regions of the Middle East and North Africa, a stream bed or channel that only carries water during the rainy season. In the southwest United States, the equivalent terms would be arroyo or wash. Wadi Hydrology has emerged as a distinct scientific area within the last decade, due mainly to the initiative of a small number of individuals within and without the Arab region, and the active support of UNESCO, assisted by ACSAD and ALECSO. This has been due to the recognition that the hydrology of arid and semi-arid areas is very different from that of humid areas and raises important scientific, technical and logistical challenges, and that an improved science base is essential to meet current and future needs of water management. By definition, water is a scarce resource in arid regions, and most countries of the Arab region and other arid and semi-arid areas are facing severe pressures due to limited water resources. Wadi bed infiltration has an important effect on flood propagation, but also provides recharge to alluvial aquifers. The balance between distributed infiltration from rainfall and wadi bed infiltration is obviously dependant on local conditions, but soil moisture observations.

Spate Hydrology

Spate hydrology is characterized by a great variation in the size and frequency of floods, which directly influence the availability of water for agriculture in any one season. Spate floods can have very high peak discharges and are usually generated in wadi catchments by localized storm rainfall. Crop production varies considerably because of the large variation in wadi runoff from year to year, season to season and day to day. Hydrological and sediment transport data are needed to design improved water diversion structures and canals in spate schemes.

Engineering Hydrology

It uses hydrologic principles in the solution of engineering problems arising from human exploitation of water resources of the earth. The engineering hydrologist, or water resources engineer, is involved in the planning, analysis, design, construction and operation of projects for the control, utilization and management of water resources.

Hydrologic calculations are estimates because mostly the empirical and approximate methods are used to describe various hydrological processes.

This branch helps in the following ways:

1. Hydrology is used to find out maximum probable flood at proposed sites e.g. Dams.

2. The variation of water production from catchments can be calculated and described by hydrology.

3. Engineering hydrology enables us to find out the relationship between a catchment's surface water and groundwater resources.

4. The expected flood flows over a spillway, at a highway Culvert, or in an urban storm drainage system can be known by this very subject.

5. It helps us to know the required reservoir capacity to assure adequate water for irrigation or municipal water supply in droughts condition.

6. It tells us what hydrologic hardware (e.g. rain gauges, stream gauges etc) and software (computer models) are needed for real-time flood forecasting.

Hydraulics

The branch on Hydraulics deals with the study of fluids: their behavior, motion and interaction of fluids with other bodies. Technically fluids include liquids and gases, but from the perspective of Hydraulics in Civil Engineering the term fluid generally means a liquid and water in particular. The topics covered include channel versus pipe flow, continuity equation, pressure distribution, specific energy, specific force, trapezoidal and circular channels, hydraulic exponent, measuring flumes, critical depth flumes, weir introduction, control structures, proportional weirs, flow over weir's, types of broad crested weirs, bear trap, sluice gate, brink depth, outlets, modules, canal lining, hydraulic jumps, spillways, surges, dam break analysis etc.

Hydraulic Engineering

Hydraulic Engineering covers the planning, management, design and operation of water supply and distribution systems, flood control and flood hazard mapping, hydrologic and hydraulic aspects of environmental issues, and application of remotely-sensed data to hydrologic and environmental problems. Hydraulic engineering is the application of fluid mechanics to the liquid earth.

Stochastic Hydrology

This branch of hydrology introduces the concepts of probability theory and stochastic processes with applications in hydrologic analysis and design. Modeling of hydrologic time series with specific techniques for data generation and hydrologic forecasting are dealt with.

Urban Hydrology

Urban hydrology is a science, part of land hydrology investigating the hydrological cycle, water regime and quality in urbanized territory. Urban hydrology is an applied science that will have an increasing role to play in the sustainability of human societies. Facing present growth of urban population, it is increasingly difficult to find and utilize new sources of water necessary to satisfy growing water demand.

Flood Hydrology

Flood hydrology entails the calculation of flood peaks or flood hydrographs for observed floods or for design floods for specified return periods. This branch of hydrology involves both the separate actions and the interactions of forces and parameters in geomorphology, hydraulics, hydrology, physical geography, economics, engineering and planning. The flood hydrology deals with two principal areas:

1. Determination of the upper limit or probable maximum flood potential at a dam site so that dams whose failure would result in loss of human life or widespread property damage can be designed to safely accommodate this flood without failure, and

2. Determination of more commonly occurring floods for use in the design of diversion dams; very low hazard storage dams; cross drainage facilities for the extensive canals, aqueducts, and roads associated with the delivery of water to users; and for diversion of flood waters that may occur during the construction of dams. Flood hydrology studies support the planning, design, construction, and operation of flood control projects.

There are three major technical aspects in flood hydrology. They are (1) hydrometeorology related to probable maximum precipitation determinations, (2) probable maximum flood hydrograph determinations, and (3) statistics and probabilities relating to the magnitude and frequency of flood flows.

Hydrometry

Field hydrometry involves the collection of all types of hydrological data involving water level, flow, rainfall and many other variables. The following aspects are covered under this branch:

1. Continuous Flow Measurement

2. Spot Flow Measurement

3. Water Level

4. Rainfall and Climate

5. Water Quality.

Highway Hydrology

Hydrologic methods of primary interest are frequency analysis for analyzing rainfall and ungaged data; empirical methods for peak discharge estimation; and hydrograph analysis and synthesis. The estimation of peak discharges of various recurrence intervals is one of the most common problems faced by engineers when designing for highway drainage structures. For the highway designer, the primary focus of hydrology is the water that moves on the earth's surface and in particular that part that ultimately crosses transportation arterials (i.e. highway stream crossings). A secondary interest is to provide interior drainage for roadways, median areas, and interchanges. By application of the principles and methods of modern hydrology, it is possible to obtain solutions that are functionally acceptable and form the basis for the design of highway drainage structures.

Wetland Hydrology

Wetland hydrology is the study of the movement of water in and out of the wetland ecosystem. In wetlands the presence of water is the overwhelming characteristic of the ecosystem. Wetlands are a unique hydrologic feature of the landscape. One particularly important attribute is their position as the transition zone between aquatic and terrestrial ecosystems. Wetlands share aspects of both aquatic and terrestrial environments because of this position. On one hand, most freshwater and marine aquatic environments, such as lakes, rivers, estuaries, and oceans, are characterized as having permanent water. On the other hand, terrestrial environments are generally characterized as having drier conditions, with an unsaturated zone present for most of the annual cycle. Wetlands thus occupy the transition zone between predominantly wet and dry environments. A diagnostic feature of wetlands is the proximity of the water surface (or water table below the surface) relative to the ground surface. In freshwater and marine aquatic habitats, the water surface lies well above the land surface, while in terrestrial environments it lies some distance below the root zone as a water table or zone of saturation. The shallow hydrologic environment of wetlands creates unique biogeochemical conditions that distinguish it from aquatic and terrestrial environments. Wetlands are a fundamental hydrologic landscape unit that generally form on flat areas or shallow slopes, where perennial water lies at or near the land surface, either above or below. Wetlands tend to form where surface and ground water accumulate within topographic depressions.

Wetland Hydrology deals with the general hydrologic properties that make wetlands unique, and to provide an overview of the processes that control wetland hydrologic behavior.

Tunnel Hydrology

All over the globe, the response of the tunnel discharge to rainfall was rapid and indicated that the flow had largely by-passed the soil matrix. Therefore, rainfall and surface runoff were infiltrating down to the subsoil (B2) horizon through established tunnel pathways.

Hydrosphere

Hydrosphere is discontinuous layer of water at or near Earth's surface. It includes all liquid and frozen surface waters, groundwater held in soil and rock, and atmospheric water vapour.

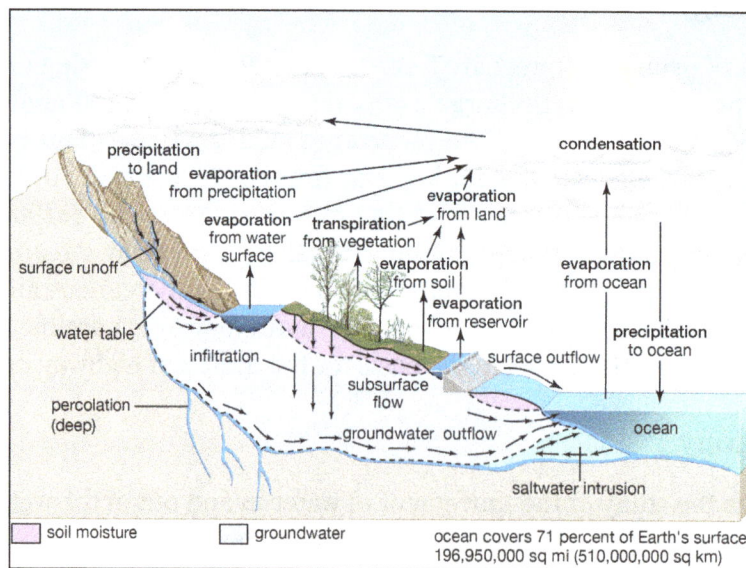

In the hydrologic cycle, water is transferred between the
land surface, the ocean, and the atmosphere.

Water is the most abundant substance at the surface of Earth. About 1.4 billion cubic km (326 million cubic miles) of water in liquid and frozen form make up the oceans, lakes, streams, glaciers, and groundwaters found there. It is this enormous volume of water, in its various manifestations, that forms the discontinuous layer, enclosing much of the terrestrial surface, known as the hydrosphere.

Central to any discussion of the hydrosphere is the concept of the water cycle (or hydrologic cycle). This cycle consists of a group of reservoirs containing water, the processes by which water is transferred from one reservoir to another (or transformed from one state to another), and the rates of transfer associated with such processes. These transfer paths penetrate the entire hydrosphere, extending upward to about 15 km (9 miles) in Earth's atmosphere and downward to depths on the order of 5 km (3 miles) in its crust.

Biogeochemical Properties of the Hydrosphere

Rainwater

About 107,000 cubic km (nearly 25,800 cubic miles) of rainfall on land each year. The total water in the atmosphere is 13,000 cubic km, and this water, owing to precipitation and evaporation, turns over every 9.6 days. Rainwater is not pure but rather contains dissolved gases and salts, fine-ground particulate material, organic substances, and even bacteria. The sources of the materials in rainwater are the oceans, soils, fertilizers, air pollution, and fossil fuel combustion.

It has been observed that rains over oceanic islands and near coasts have ratios of major dissolved constituents very close to those found in seawater. The discovery of the high salt content of rain near coastlines was somewhat surprising because sea salts are not volatile, and it might be expected that the process of evaporation of water from the sea surface would "filter" out the salts. It has been demonstrated, however, that a large percentage of the salts in rain is derived from the bursting of small bubbles at the sea surface due to the impact of rain droplets or the breaking of waves, which results in the injection of sea aerosol into the atmosphere. This sea aerosol evaporates, with resultant precipitation of the salts as tiny particles that are subsequently carried high into the atmosphere by turbulent winds. These particles may then be transported over continents to fall in rain or as dry deposition.

Assuming equilibrium with the atmospheric carbon dioxide partial pressure (P_{CO_2}) of $10^{-3.5}$ (0.00035) atmosphere, the approximate mean composition of rainwater is in parts per million (ppm): sodium (Na^+), 1.98; potassium (K^+), 0.30; magnesium (Mg^{2+}), 0.27; calcium (Ca^{2+}), 0.09; chloride (Cl^-), 3.79; sulfate (SO_4^{2-}), 0.58; and bicarbonate (HCO_3^-), 0.12. In addition to these ions, rainwater contains small amounts of dissolved silica—about 0.30 ppm. The average pH value of rainwater is 5.6. (The term pH is defined as the negative logarithm of the hydrogen ion concentration in moles per litre. The pH scale ranges from 0 to 14, with lower numbers indicating increased acidity). On a global basis, as much as 35 percent of the sodium, 55 percent of the chlorine, 15 percent of the potassium, and 37 percent of the sulfate in river water may be derived from the oceans through sea aerosol generation.

A considerable amount of data has become available for marine aerosols. These aerosols are important because (1) they are vital to any description of the global biogeochemical cycle of an element, (2) they may have an impact on climate, (3) they are a sink, via heterogeneous chemical reactions, for trace atmospheric gases, and (4) they influence precipitation of cloud and rain droplets. For many trace metals, the ratio of the atmospheric flux to the riverine flux for coastal and remote oceanic areas may be greater than one, indicating the importance of atmospheric transport. Figures have been prepared that illustrate the enrichment factors (EF) of North Atlantic marine aerosols and suspended matter in North Atlantic waters relative to the crust (that is, terrestrial sources), where

$$EF_{crust} = \frac{(X / Al)_{air}}{(X / Al)_{crust}},$$

and $(X/Al)_{air}$ and $(X/Al)_{crust}$ refer, respectively, to the ratio of the concentration of the element X to that of Al, aluminum (which is an easily observed terrestrial component of aerosols), in the atmosphere and in average crustal material. Comparing the enrichment factors in marine aerosols with those of suspended matter in the water column indicates qualitatively the marine aerosols' importance as a source that alters the composition of marine suspended matter and, consequently, their importance to deep-sea sedimentation. Moreover, such comparisons help identify how significant terrestrial sources are for both the marine aerosols and the water below.

In some instances the ratios of ions in rainwater deviate significantly from those in seawater. Mechanisms proposed for this fractionation are, for example, the escape of chlorine as gaseous hydrogen chloride (HCl) from sea salt aerosol with a consequent enrichment in sodium and bubbling and thermal diffusion. In addition, release of biogenic gases such as dimethyl sulfide (DMS) from the sea surface and its subsequent reaction in the oceanic atmosphere to sulfate can change rainwater ion ratios with respect to seawater. Soil particles also can influence rainwater composition. Rainfall over the southwestern United States contains relatively high sulfate concentrations because of sulfate-bearing particles that have been blown into the atmosphere from desert soils. Rain near industrial areas commonly contains high contents of sulfate, nitrate, and carbon dioxide (CO_2) largely derived from the burning of coal and oil. There are two main processes leading to the conversion of sulfur dioxide (SO_2) to sulfuric acid (H_2SO_4). These are reactions with hydroxyl radicals ($-OH$) and with hydrogen peroxide (H_2O_2) in the atmosphere:

$$SO_2 + OH \rightarrow \text{intermediate species} \rightarrow H_2SO_4$$

and

$$SO_2 + H_2O_2 = H_2SO_4.$$

The sulfuric acid then dissociates to hydrogen and sulfate ions:

$$H_2SO_4 = 2H^+ + SO_4^{2-}.$$

For the nitrogen gases nitric oxide (NO) and nitrogen dioxide (NO_2) released from fossil fuel burning, their atmospheric reactions lead to the production of nitric acid (HNO_3) and its dissociation to hydrogen ions (H^+) and nitrate (NO_3^-). These reactions are responsible for the acid rain conditions that occurred in the northeastern United States, southeastern Canada, and western Europe during the second half of the 20th century. The high sulfate values of the rain in the northeastern United States reflect the acid precipitation conditions of this region.

River and Ocean Waters

River discharge constitutes the main source for the oceans. Seawater has a more uniform composition than river water. It contains, by weight, about 3.5 percent dissolved salts, whereas river water has only 0.012 percent. The average density of the world's oceans is roughly 2.75 percent greater than that of typical river water. Of the average 35 parts per thousand salts of seawater, sodium and chlorine make up almost 30 parts, and magnesium and sulfate contribute another four parts. Of the remaining one part of the salinity, calcium and potassium constitute 0.4 part each and carbon, as carbonate and bicarbonate, about 0.15 part. Thus, nine elements (hydrogen, oxygen, sulfur, chlorine, sodium, magnesium, calcium, potassium, and carbon) make up 99 percent of seawater, though most of the 94 naturally occurring elements have been detected therein. Of importance are the nutrient elements phosphorus, nitrogen, and silicon, along with such essential micronutrient trace elements as iron, cobalt, and copper. These elements strongly regulate the organic production of the world's oceans.

A portion of the delta of the Mekong River as it flows through
southern Vietnam and empties into the South China Sea.

In contrast to ocean water, the average salinity of the world's rivers is low—only about 0.012 percent, or 120 ppm by weight. Of this salt content, carbon as bicarbonate constitutes 58 parts, or 48 percent, and calcium, sulfur as sulfate, and silicon as dissolved monomeric silicic acid make up a total of about 39 parts, or 33 percent. The remaining 19 percent consists predominantly of chlorine, sodium, and magnesium in descending importance. It is obvious that the concentrations and relative proportions of dissolved species in river waters contrast sharply with those of seawater. Thus, even though seawater is derived in part by the chemical differentiation and evaporation of river water, the processes involved affect every element differently, indicating that simple evaporation and concentration are entirely secondary to other processes.

Water-rock Interactions as Determining River Water Composition

Generally speaking, the composition of river water, and thus that of lakes, is controlled by water-rock interactions. The attack of carbon dioxide-charged rain and soil waters on the individual minerals in continental rocks leads to the production of dissolved constituents for lakes, rivers, and streams. It also gives rise to solid alteration products that make up soils or suspended particles in freshwater aquatic systems. The carbon dioxide content of rain and soil waters is of particular importance in weathering processes. The pH of rainwater equilibrated with the atmospheric carbon dioxide partial pressure of $10^{-3.5}$ (0.00032) atmosphere is 5.6. In industrial regions, rainwater pH values may be lower because of the release and subsequent hydrolysis of acid gases—namely, sulfur dioxide and nitrogen oxides (NO_x) from the combustion of fossil fuels. After rainwater enters soils, its characteristics change markedly. The usual few part per million of salts in rainwater increase substantially as the water reacts. The upper part of the soil is a zone of intense biochemical activity. The bacterial population near the surface is large, but it decreases rapidly downward. One of the major biochemical processes of the bacteria is the oxidation of organic material, which leads to the release of carbon dioxide. Soil gases obtained above the zone of water saturation may contain 10 to 40 times as much carbon dioxide as the free atmosphere, and in some cases carbon dioxide has been shown to make up 30 percent of the soil gases as opposed to 0.03 percent of the free atmosphere. In addition to the acid effects of carbon dioxide, there is a highly acidic microenvironment created by the roots of living plants. Values of pH as low as 2 have been measured immediately adjacent to root hairs. The combined length of a plant's root hairs may be several kilometres, so their chemical effects on acidic water are formidable.

Suspended soil particles: The Amazon River near Manaus, Brazil. The brown colour of the water is the result of rains washing soil particles into the stream from surrounding land and the stirring up of mud on the riverbed.

Congruent and Incongruent Weathering Reactions

These acid solutions in the soil environment attack the rock minerals, the bases of the system, producing neutralization products of dissolved constituents and solid particles. Two general types of reactions occur: congruent and incongruent. In the former, a solid dissolves, adding elements to the water according to their proportions in the mineral. An example of such a weathering reaction is the solution of calcite ($CaCO_3$) in limestones:

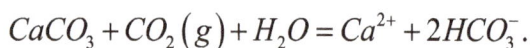

$$CaCO_3 + CO_2(g) + H_2O = Ca^{2+} + 2HCO_3^-.$$

Here one of the HCO_3^- ions comes from calcite and the other from $CO_2(g)$ in the reacting water. The amount of carbon dioxide dissolved according to reaction $CaCO_3 + CO_2(g) + H_2O = Ca^{2+} + 2HCO_3^-$ depends on temperature, pressure, original bicarbonate content of the weathering solution, and partial pressure of the carbon dioxide. The carbon dioxide and the temperature are the most important variables. Increases in one or both of these variables lead to increases in the amount of calcite dissolved. For example, for a carbon dioxide pressure of $10^{-3.5}$ (0.00032) atmosphere, the amount of calcium that can be dissolved until saturation is about $10^{-3.3}$ (0.0005) mole, or 20 ppm, at 25 °C (77 °F). For an atmospheric carbon dioxide pressure of 10^{-2} (0.01) atmosphere and for a soil atmosphere of nearly pure carbon dioxide, the values are 65 and 300 ppm, respectively. The weathering of calcite leads to the release of calcium and bicarbonate ions into soil waters and groundwaters, and these constituents eventually reach lake and river systems. The insoluble residue of quartz (SiO_2), clay minerals, and iron oxides (e.g. FeOOH) in the limestone rock make up the deep-red soils that form from limestone weathering. These particles may be carried into streams by runoff and hence to lakes and the oceans and become part of the suspended load of these systems.

An example of an incongruent weathering reaction—that is, one where only part of a solid is consumed—is that involving aluminosilicates. One such reaction is the aggressive attack of carbon dioxide-charged soil water on the mineral K-spar ($KAlSi_3O_8$), an important phase found in continental rocks. The reaction is:

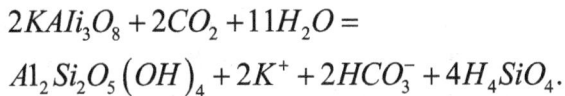

$$2KAli_3O_8 + 2CO_2 + 11H_2O =$$
$$Al_2Si_2O_5(OH)_4 + 2K^+ + 2HCO_3^- + 4H_4SiO_4.$$

It should be noted that the K-spar changes into a new mineral—kaolinite (a clay mineral) in this case—plus solution, and acid is consumed. The total dissolved material per litre of soil solution released is about 60 ppm for a solution initially equilibrated with a typical soil carbon dioxide content.

The water resulting from reaction:

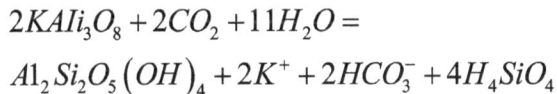

$$2KAli_3O_8 + 2CO_2 + 11H_2O =$$
$$Al_2Si_2O_5(OH)_4 + 2K^+ + 2HCO_3^- + 4H_4SiO_4$$

would contain bicarbonate, potassium, and dissolved silica in the ratios 1:1:2, and the new solid, kaolinite, would be a weathering product. These dissolved constituents and the solid alteration product would eventually reach rivers to be transferred possibly to lakes and ultimately to the sea. It has been demonstrated that the composition of river water is the product of a variety of mineral-water reactions such as:

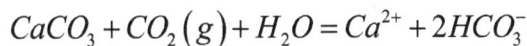

$$CaCO_3 + CO_2(g) + H_2O = Ca^{2+} + 2HCO_3^-$$

and

$$2KAli_3O_8 + 2CO_2 + 11H_2O =$$
$$Al_2Si_2O_5(OH)_4 + 2K^+ + 2HCO_3^- + 4H_4SiO_4 .$$

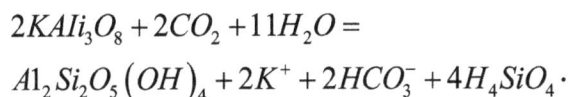

The dissolved load of the world's rivers comes from the following sources: 7 percent from beds of halite (NaCl) and salt disseminated in rocks, 10 percent from gypsum ($CaSO_4 \cdot 2H_2O$) and anhydrite ($CaSO_4$) deposits and sulfate salts disseminated in rocks, 38 percent from limestones and dolomites, and 45 percent from the weathering of one silicate mineral to another. Of the bicarbonate ions in river water, 56 percent stems from the atmosphere, 35 percent from carbonate minerals, and 9 percent from the oxidative weathering of fossil organic matter. Reactions involving silicate minerals account for 30 percent of the riverine bicarbonate ions.

Besides dissolved substances, rivers also transport solids in traction (i.e. bed load) and, most importantly, suspended load. The present global river-borne flux of solids to the oceans is estimated as 15.5 billion metric tons (about 17.1 billion tons) per year. Present elemental fluxes are estimated in millions of metric tons per year as silicon, 4,420; aluminum, 1,460; iron, 740; calcium, 330; potassium, 310; magnesium, 210; and sodium, 110. The total load of particulate organic carbon of the world's rivers is 180 million metric tons (nearly 200 million tons) per year. The riverine fluxes of trace metals to the oceans are dominated by their occurrence in the particulate phase as opposed to the dissolved phase. The particulate matter in river water is an important source of silicon, aluminum, iron, titanium, rubidium, scandium, vanadium, the lanthanoids, and other elements for deep-sea sediments.

Lake Waters

Although lake waters constitute only a small percentage of the water in the hydrosphere, they are an important ephemeral storage reservoir for fresh water. Aside from their recreational use, lakes constitute a source of water for household, agricultural, and industrial uses. Lake waters are also very susceptible to changes in chemical composition due to these uses and to other factors.

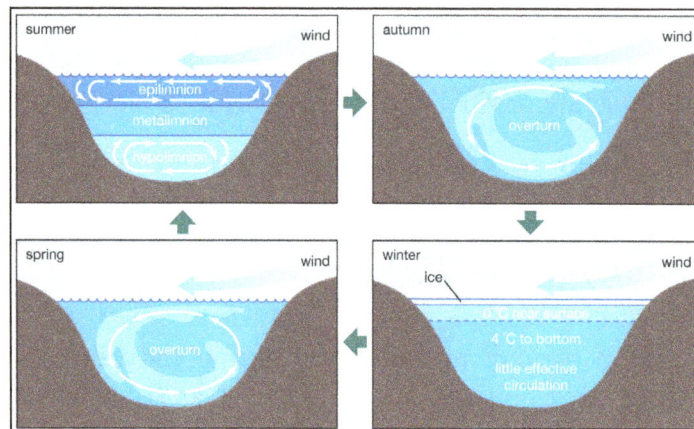

Temperate lake circulation

The above image shows the Annual circulation patterns in a dimictic lake. The typical dimictic lake has distinct layers that fully mix twice a year. It undergoes stratification in the summer and complete overturn in the autumn and spring. During winter, surface ice prevents further mixing by the wind. Small differences in density and temperature exist, with cooler water (0 °C [32 °F]) staying near the surface and warmer, denser water (4 °C [39.2 °F]) extending to the bottom.

In general, fresh waters at the continental surface evolve from their rock sources by enrichment in calcium and sodium and by depletion in magnesium and potassium. In very soft waters the alkalies may be more abundant than the alkaline earths, and in the more-concentrated waters of open river systems calcium > magnesium > sodium > potassium. For the anions, in general, HCO_3^- exceeds SO_4^{2-} which is greater in concentration than Cl^-. It is worthwhile at this stage to consider some major mechanisms that control global surface water composition. These mechanisms are atmospheric precipitation, rock reactions, and evaporation-precipitation.

The mechanism principally responsible for waters of very low salinity is precipitation. These waters tend to form in tropical regions of low relief and thoroughly leached source rocks. In these regions rainfall is high, and volumes of fresh water (rivers, tributary streams, pools, etc.) within a watershed are usually dominated by salts brought in by precipitation. Such waters constitute one part of a chain of water volumes that begins with falling precipitation and ends with the release of water into the ocean, for which the final part of the chain represents water volumes dominated by contributions of dissolved salts from the rocks and soils of their basins. These waters have moderate salinity and are rich in dissolved calcium and bicarbonate. They are in turn the "end-member" of another series that extends from the calcium-rich, medium-salinity fresh waters to the high-salinity, sodium chloride-dominated waters of which seawater is an example. Seawater composition, however, does not evolve directly from the composition of fresh waters and the precipitation of calcium carbonate; other mechanisms that control its composition are involved. Such factors as relief and vegetation also may affect the composition of the world's surface waters, but atmospheric precipitation, water-rock reactions, and evaporation-crystallization processes appear to be the dominant mechanisms governing continental surface water chemistry.

Continental fresh waters evaporate once they have entered closed basins, and their constituent salts precipitate on the basin floors. The composition of these waters may evolve along several different paths, depending on their initial chemical makeup.

Biological processes strongly affect the composition of lake waters and are responsible to a significant degree for the compositional differences between the upper water layer (the epilimnion) and the lower water layer (the hypolimnion) of lakes. The starting point is photosynthesis, represented by the following reaction:

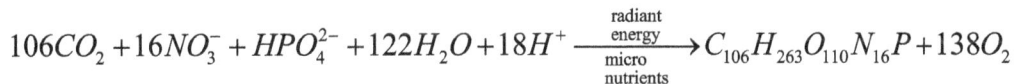

$$106CO_2 + 16NO_3^- + HPO_4^{2-} + 122H_2O + 18H^+ \xrightarrow[\substack{micro \\ nutrients}]{\substack{radiant \\ energy}} C_{106}H_{263}O_{110}N_{16}P + 138O_2$$

The reversal of this reaction is oxidation-respiration leading to the release of the nutrients nitrogen and phosphorus, as well as carbon dioxide. In a stratified lake, carbon, nutrients, and silica are extracted from the upper layer during photosynthesis. This process leads to reduced concentrations of nitrate, phosphate, and silica in these waters and, during times of maximum daylight organic production, to supersaturation of the upper layer with respect to dissolved oxygen. The organic matter produced by phytoplankton may be either grazed upon by zooplankton and other organisms or decomposed by bacteria. Some of it, however, sinks into the lower layer. There it is further decomposed, especially by bacteria, resulting in the release of dissolved phosphorus and nitrogen and the consumption of oxygen. Oxygen concentrations therefore are reduced in these lower lake waters, because stratification prevents oxygen exchange with the atmosphere. Furthermore, the inorganic carbonate and siliceous skeletons of the dead organisms sinking into the lower layer may dissolve, giving rise to increased concentrations of dissolved silica and inorganic carbon in the deep waters of stratified lakes. This dissolution is a result of undersaturation of the waters of the lower layer with respect to the opaline silica and calcium carbonate that make up the skeletons of the dead and sinking plankton. These natural biological processes have been accelerated in some lakes because of excess nutrient input by human activity, resulting in the eutrophication of lake waters and marine systems.

Groundwaters

Groundwaters derive their compositions from a variety of processes, including dissolution, hydrolysis, and precipitation reactions; adsorption and ion exchange; oxidation and reduction; gas exchange between groundwater and the atmosphere; and biological processes.

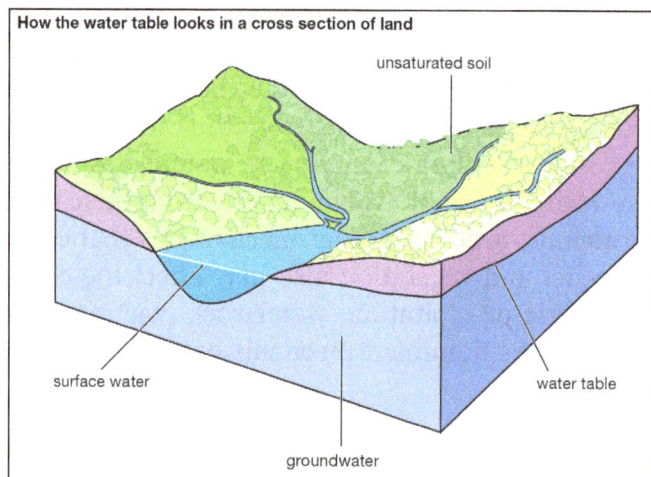

The water table is the top level of groundwater. Surface water is an exposed part of the water table.

The biological processes of greatest importance are microbial metabolism, organic production, and respiration (oxidation). By far the most important overall process for the major constituents of groundwater is that of mineral-water reactions. Thus, the composition of groundwaters strongly reflects the types of rock minerals that the waters have encountered in their movement through the subsurface.

In general, the most mobile elements in groundwater—i.e. those most easily liberated by the weathering of rock minerals—are calcium, sodium, and magnesium. Silicon and potassium have intermediate mobilities, and aluminum and iron are essentially immobile and locked up in solid phases.

Groundwaters are highly susceptible to contamination because of human activities and the fact that their dissolved constituents are derived to a large extent from the leaching of surface materials. Some of the nitrogen and phosphorus applied to soils as fertilizers and organic pesticides may be leached and leak into groundwater systems, leading to increased concentrations of ammonium and phosphate. Radioactive wastes, industrial chemicals, household materials, and mine refuse are other anthropogenic sources of dissolved substances that have been detected in groundwater systems.

Ice

Ice is nearly a pure solid and, as such, accommodates few foreign ions in its structure. It does contain, however, particulate matter and gases, which are trapped in bubbles within the ice. The change in composition of these materials through time, as recorded in the successive layers of ice, has been used to interpret the history of Earth's surface environment and the impact of human activities on this environment. The increase in the lead content of continental glacial ice with decreasing age of the ice up to the middle of the 1970s, for example, reflects the progressive input of tetraethyl lead into the global environment from gasoline burning. (Stringent environmental regulations that appeared in the 1970s regarding the use of leaded gasoline has led to a fall in lead concentrations in ice laid down since that time.) Also, atmospheric carbon dioxide and methane concentrations, which have increased significantly during the past century because of anthropogenic activities, are faithfully recorded in ice bubbles of the thick continental ice sheets. By 2016 atmospheric carbon dioxide and methane concentrations had increased by more than 43 percent and more than 150 percent, respectively, higher than their concentrations 200 years ago; the latter concentration values were obtained from measurements of the gases in air trapped in ice.

Perito Moreno Glacier, Los Glaciares National Park: Perito Moreno Glacier, Los Glaciares National Park, Argentina. Although many other South American glaciers have declined in mass since the middle of the 20th century, the mass of Perito Moreno Glacier has remained relatively steady.

Importance of Hydrosphere

The hydrosphere is of great importance as it plays an integral role in the survival of all life forms. Here are some of the significant functions of the hydrosphere on Earth:

1. A Component of Living Cells: Each cell in a living organism is composed of at least 75% water. This promotes the normal functioning of the cell. Most of the chemical reactions that occur in living organisms involve materials that are dissolved in water. No cell would survive or be able to carry its normal functions without water.

2. Habitat for Many Life Forms: The hydrosphere provides an important place for a wide range of plants and animals to live. Many nutrients such as nitrate, nitrite, and ammonium ions, as well as gases such as carbon dioxide and oxygen are dissolved in water. These compounds play an integral role in the existence of life in water.

3. Climate Regulation: One of water's exceptional features is its high specific heat. Namely, water takes not only a long time to heat up but also a long time to cool down. You know what's the significance of this? It plays a significant role in regulating temperatures on earth, ensuring temperatures remain within a range that is suitable for the existence of life. Ocean currents also play a critical role in heat dispersion.

4. Existence of Atmosphere: Hydrosphere has a significant contribution to the existence of the atmosphere in its present form. When the earth was formed it comprised only a very thin atmosphere. This atmosphere was packed with helium and hydrogen similar to Mercury's current atmosphere.

   The gases helium and hydrogen were later ejected from the atmosphere. And the gases and water vapor produced as the Earth cooled became its current atmosphere. The volcanoes also released other gases and water vapor, which entered the atmosphere.

   This process is estimated to have occurred about 400 million years ago.

5. Human Needs: The hydrosphere benefits humans in numerous ways. Besides drinking, water is used for domestic purposes like cooking and cleaning as well as for industrial purposes. Water can also be used for transportation, agriculture, and to generate electricity through hydropower.

Impact of Human Activities on the Hydrosphere

The activities of modern society are having a severe impact on the hydrologic cycle. The dynamic steady state is being disturbed by the discharge of toxic chemicals, radioactive substances, and other industrial wastes and by the seepage of mineral fertilizers, herbicides, and pesticides into surface and subsurface aquatic systems. Inadvertent and deliberate discharge of petroleum, improper sewage disposal, and thermal pollution also are seriously affecting the quality of the hydrosphere.

Eutrophication

Historically, aquatic systems have been classified as oligotrophic or eutrophic. Oligotrophic waters are poorly fed by the nutrients nitrogen and phosphorus and have low concentrations of these constituents. There is thus low production of organic matter by photosynthesis in such waters. By contrast, eutrophic waters are well supplied with nutrients and generally have high concentrations of nitrogen and phosphorus and, correspondingly, large concentrations of plankton owing to high biological productivity. The waters of such aquatic systems are usually murky, and lakes and coastal marine systems may be oxygen-depleted at depth. The process of eutrophication is defined as high biological productivity resulting from increased input of nutrients or organic matter into aquatic systems. For lakes, this increased biological productivity usually leads to decreased lake volume because of the accumulation of organic detritus. Natural eutrophication occurs as aquatic systems fill in with organic matter; it is distinct from cultural eutrophication, which is caused by human intervention. The latter is characteristic of aquatic systems that have been artificially enriched by excess nutrients and organic matter from sewage, agriculture, and industry. Naturally eutrophic lakes may produce 75–250 grams of carbon per square metre per year, whereas those lakes experiencing eutrophication because of human activities can support 75–750 grams per square metre per year. Commonly, culturally eutrophic aquatic systems may exhibit extremely low oxygen concentrations in bottom waters. This is particularly true of stratified systems, as, for instance, lakes during summer where concentrations of molecular oxygen may reach levels of less than about one milligram per litre—a threshold for various biological and chemical processes.

Aquatic systems may change from oligotrophic to eutrophic, or the rate of eutrophication of a natural eutrophic system may be accelerated by the addition of nutrients and organic matter due to human activities. The process of cultural eutrophication, however, can be reversed if the excess nutrient and organic matter supply is shut off.

Not only do freshwater aquatic systems undergo cultural eutrophication, but coastal marine systems also may be affected by this process. On a global scale, the input by rivers of organic matter to the oceans today is twice the input in prehuman times, and the flux of nitrogen, together with that of phosphorus, has more than doubled. This excess loading of carbon, nitrogen, and phosphorus is leading to cultural eutrophication of marine systems. In several polluted eastern U.S. estuaries (e.g. Chesapeake and Delaware bays) and in some estuaries of western Europe (e.g. the Scheldt of Belgium and the Netherlands), all of the dissolved silica brought into the estuarine waters by rivers is removed by phytoplankton growth (primarily diatoms) resulting from excess fluxes of nutrients and organic matter. In the North Sea there is now a deficiency of silica and an excess of nitrogen and phosphorus, which in turn has led to a decrease in diatom productivity and an increase in cyanobacteria productivity—a biotic change brought about by cultural eutrophication.

Acid Rain

The emission of sulfur dioxide and nitrogen oxides to the atmosphere by human activities—primarily fossil-fuel burning—has led to the acidification of rain and freshwater aquatic systems. Acid rain is a worldwide problem and has been well documented for eastern North America and the countries of western Europe.

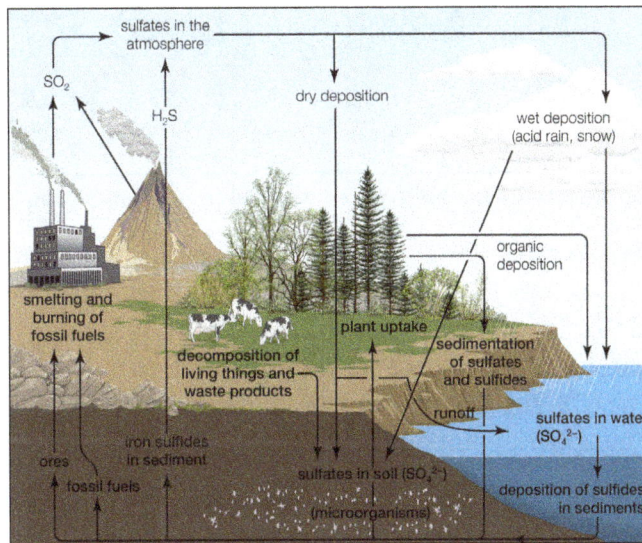

Sulfur cycle: Major sulfur-producing sources include sedimentary rocks, which release hydrogen sulfide gas, and human sources, such as smelters and fossil-fuel combustion, both of which release sulfur dioxide into the atmosphere.

Acid rain is defined as precipitation with a pH of less than 5.2 that results from reactions involving gases other than carbon dioxide.

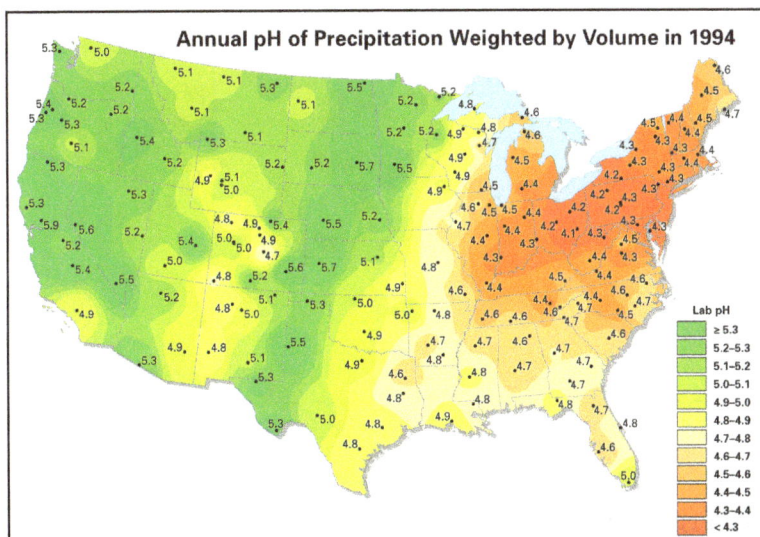

Map of precipitation pH in the continental United States in 1994.

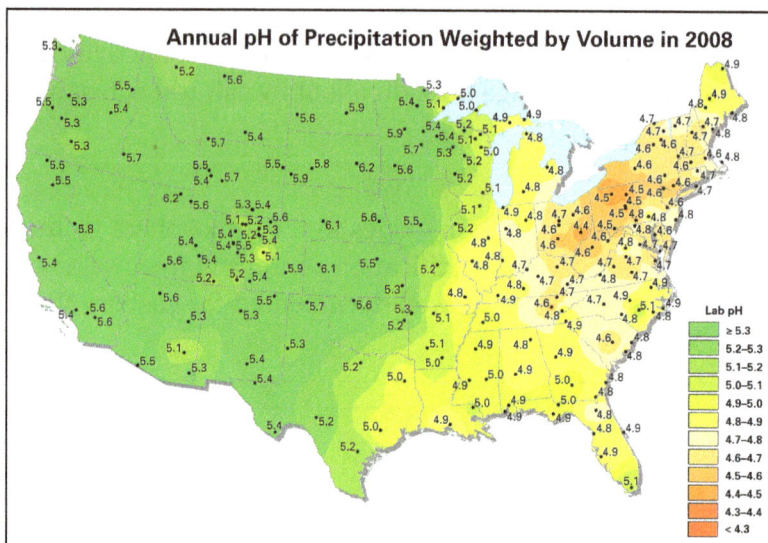

Map of precipitation pH in the continental United States in 2008.

Average pH can be calculated as the $-\log a_H^+$ (a_H^+ is activity of the hydrogen ion). A water chemistry study examining precipitation over the eastern United States for the period October 1979 through September 1980 revealed that low pH values were a result of equilibration of rainwater with the atmospheric acid gases of carbon, nitrogen, and sulfur. Equilibration only with atmospheric carbon dioxide would give a pH of 5.7. The significantly lower values are a result of reactions with nitrogen- and sulfur-bearing gaseous atmospheric components derived primarily from fossil-fuel burning sources.

Nitrate and sulfate concentrations in precipitation over the eastern United States are strongly correlated with pH—the lower the pH of rain, the higher the concentrations of nitrate and sulfate. Such low pH values and increased nitrate and sulfate concentrations were observed in the rains of western Europe and North America until the late 20th century. The pH values of precipitation in these regions have increased significantly since then because of strict air quality regulations. Other parts of the world that have industrialized since the late 20th century without enacting adequate air pollution controls, such as China, experienced similar pH declines in precipitation.

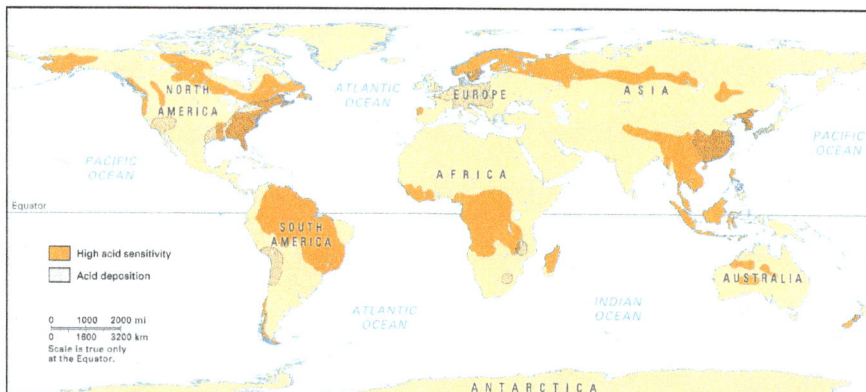

Areas affected by acid deposition contrasted with regions of high acid sensitivity.

Wet and dry deposition also removes the hydrogen ion produced in the rain by the oxidation and hydrolysis of these acid gases. This excess hydrogen ion can bring about the acidification of freshwater aquatic systems, particularly those with little buffer capacity (e.g. lakes situated in crystalline rock terrains). Furthermore, the lower pH values of rainwater, and consequently of soil water, can lead to increased mobilization of aluminum. Acidification of freshwater lakes in the eastern United States and southeastern Canada and increased aluminum concentrations in their waters are thought to be responsible for major changes in the ecosystems of the lakes. In particular, many lakes of this region lack substantial fish populations today, even though they supported large numbers of fish in the early 1900s. Acid rain also may be among the factors responsible for damage to the major forests and soils of the eastern United States and western Europe.

Buildup of Greenhouse Gases

One problem that was brought about by human action and is definitely affecting the hydrosphere globally is that of the greenhouse gases (so called because of their heat-trapping "greenhouse" properties) emitted to the atmosphere. Of the greenhouse gases released by anthropogenic activities, carbon dioxide has received much attention. Measurements of carbon dioxide in air bubbles trapped in ice and the continuous measurement of carbon dioxide concentrations in air samples collected at Mauna Loa, Hawaii, since 1958 show that the atmospheric concentration of more than 400 ppmv is roughly 45 percent higher than its late 1700s value of 275 ppmv. Much of this increase is due to carbon dioxide released to the atmosphere from the burning of coal, oil, gas, and wood and from the slash-and-burn activities that accompany deforestation practices (as, for example, those adopted in the Amazon River basin). The component of the hydrosphere most greatly affected by this emission of carbon dioxide is the ocean.

Before human activities had substantially affected the carbon dioxide cycle, there was a net flux of carbon dioxide from the oceans through the atmosphere to the land, where the gas was used in the net production of organic matter and the chemical weathering of minerals in continental rocks. Because of fossil-fuel burning and land-use practices, the net transfer from the ocean to the land has been reversed, and the ocean has now become an important sink of carbon dioxide. The net chemical reaction of adding carbon dioxide to the ocean (provided there is no reaction with carbonate solids) is:

$$CO_2 + H_2O + CO_3^{2-} \left(\text{carbonate ion}\right) = 2HCO_3^-$$

and a lowering of the pH of surface seawater. Such a pH effect has not been observed but conceivably could occur if carbon dioxide continues to be released to the atmosphere by human activities.

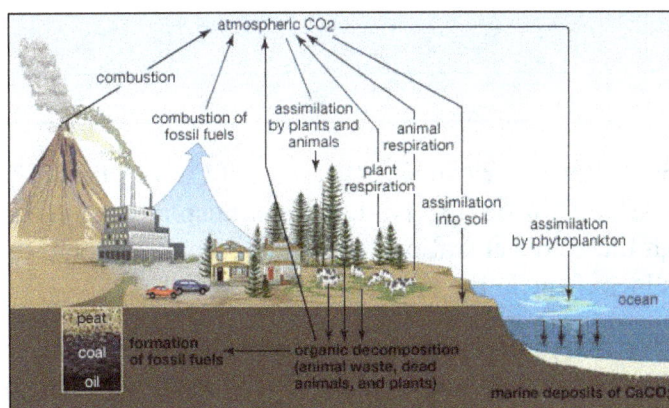

The generalized carbon cycle

Based on greenhouse climate models and other considerations, it is possible that atmospheric carbon dioxide concentrations may double from their late 1700s level of 275 ppmv by the years 2030–50. By 2015, atmospheric carbon dioxide concentrations had surpassed 400 ppmv for the first time in 800,000 years. Climate models, which also consider the long-term warming potential made by other greenhouse gases (e.g. methane and nitrous oxide) in addition to that of carbon dioxide, project a rise in global mean surface temperature of 0.3 to 4.8 °C (0.5 to 8.6 °F) by 2100. This projected temperature increase would be two to three times greater at the poles than at the Equator and greater in the Arctic than in the Antarctic. At present there is no systematic world-wide program to decrease greenhouse gas emissions, except for that affecting chlorofluorocarbon (Freon) releases. Thus, it is conceivable that atmospheric carbon dioxide concentrations in the late 21st and early 22nd centuries might reach levels greater than twice their 1700s value. Whatever the case, the effect of the potential rise in surface temperature would be to speed up the hydrologic cycle and probably the rate of chemical weathering of continental rocks. Increases in the global mean evaporation and precipitation rates are expected from a doubling of the carbon dioxide level and a few degrees rise in global mean temperature. The effect on the water balance would be regional in nature, with some places becoming wetter and others drier. In general, there would be a trend toward greater and longer periods of summer dryness induced by lower soil moisture content and higher evaporation rates in the mid-latitudes of the Northern Hemisphere. In the arid western regions of the United States, which depend on irrigation for growing plants, severe water shortages could occur. By contrast, precipitation and runoff might increase, except in summer, at latitudes beyond 60° N because of a greater poleward transport of moisture. In summer, in a zone centred around 60° N, greater dryness might occur as a result of an earlier end of snowmelt.

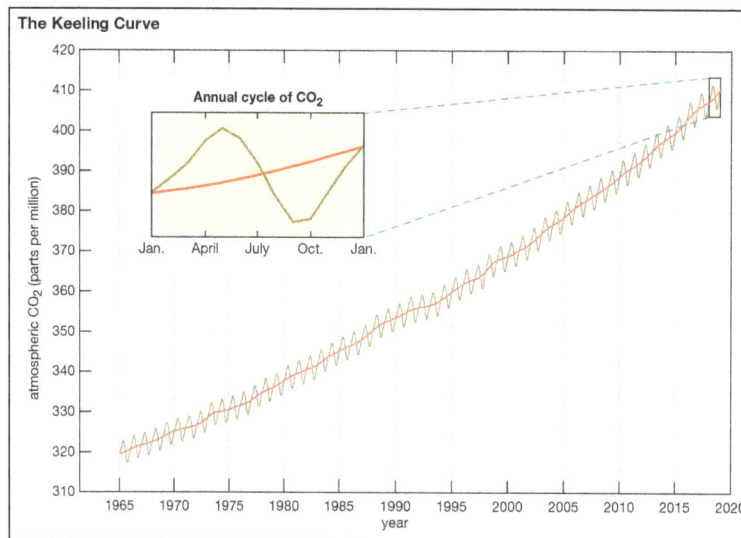

The Keeling Curve, named after American climate scientist Charles David Keeling, tracks changes in the concentration of carbon dioxide (CO_2) in Earth's atmosphere at a research station on Mauna Loa in Hawaii. Although these concentrations experience small seasonal fluctuations, the overall trend shows that CO_2 is increasing in the atmosphere.

Global warming could further affect the hydrologic cycle by the melting of ice and snow in the Greenland and Antarctic ice caps and in mountain glaciers, resulting in the transfer of water to the oceans. This process, together with thermal expansion of the oceans because of global warming, resulted in a slow rise in sea level during the 20th century, and it is expected to continue throughout the 21st century. If the West Antarctic Ice Sheet were to disintegrate, a much larger and more rapid rise in sea level of more than 3 metres (nearly 10 feet) could occur over the next several hundred years. The melting of all glacial ice would raise the sea level more than 66 metres (about 216 feet).

Significant reductions in the areal extent and thickness of sea ice in the Arctic occurring during the early 21st century have been attributed to global warming. Complete melting of the Arctic sea ice might occur, causing a northward shift in storm tracks and a reduction in Northern Hemispheric precipitation during the spring and fall. Furthermore, a worldwide reduction in sea ice might lead to increased evaporation from the ocean and increased low-altitude cloudiness, which would reflect solar radiation and cause cooling.

The potential changes in the hydrologic cycle induced by global warming resulting from anthropogenic emissions of greenhouse gases do not seem great. Yet, their consequences could be severe for ecosystems and human populations, especially since the latter are so sensitive to and dependent on such changes. A global rise in sea level of 1 metre (3.3 feet), for example, would almost completely inundate the coastal areas of Bangladesh. Agricultural lands could be displaced, just as patterns of arid, semiarid, and wet lands might become modified. It is essential that society plan for such potential changes so that, if they do occur, appropriate adjustments can be made to ameliorate them.

Cryosphere

The cryosphere is the frozen part of Earth's surface. This includes glaciers, snow, sea ice, freshwater ice, permafrost, ice caps, and ice sheets. The cryosphere plays a significant role in the global climate by influencing hydrology, clouds, precipitation, and circulation of the atmosphere and the oceans. The term cryology refers to the science of cryospheres and deglaciation is the loss of the cryosphere (such as the reduction of global ice content due to global warming).

Components

Ice and snow are the main part of any component of the cryosphere, but there is an extensive list of formations that make up the cryosphere. These include:

- Land-based Ice: Land-based ice makes up the largest parts of the cryosphere. Occasionally, land-based or continental ice can flow from land to the sea. When this happens, you get shelf ice. Some components of this part of the cryosphere include:

    - Continental ice sheets: Large masses of ice on land in Greenland and Antarctica.

    - Ice caps.

    - Glaciers: Glaciers are large masses of ice on land that have built up from many seasons of snowfall. Slowly, glaciers move downhill. Glaciers cover about 10% of the Earth's surface. They also act as a large place of water storage for fresh water.

    - Permafrost: Frozen soil or rock wherein almost all of the water held within it has frozen. If the ground is frozen all year, it is called permafrost.

- Land-based Snow: Snow is a form of precipitation that falls in solid crystals. Snow is found in places all over the world and is important for some plants and animals.

- Water-based Ice: This is frozen parts of otherwise liquid components of the hydrosphere. Much of the water-based ice around the world is around Antarctica and the Arctic. Components of this part of the cryosphere are:

  ◦ Icebergs: These are chunks of ice floating in water that break off of glaciers and ice shelves.

  ◦ Ice shelves: Platforms of ice that form where ice sheets and glaciers move into oceans.

  ◦ Sea ice: This is ice that forms in the ocean when water is cooled to temperatures below freezing. Sea ice is located primarily in Arctic and Antarctic Oceans.

  ◦ Frozen rivers and lakes.

Physical Properties of the Cryosphere

The existence of the cryosphere varies widely depending on its specific location around the world. For example, snow and freshwater ice may exist only through winter seasons in many places, whereas many glaciers have been frozen for over 10,000 years. Antarctica is home to the majority of global ice volume however, the northern hemisphere is home to the largest cryosphere area. Climate researchers rely on measurements of the cryosphere for information about global climate change.

The cryosphere affects the world's climate via three distinct properties: surface reflectance, thermal diffusivity, and latent heat.

Surface Reflectance

Much of the cryosphere reflects the sun's solar radiation; this reaction is known as surface reflectance. Surface reflectance is measured by the difference between reflected and incident solar radiation, known as albedo. In other words, albedo is the reflecting power of a particular surface. Some of the highests rates of albedo, between 80% and 90%, are found in areas with year-round snow coverage.

During autumn and spring, the rates of albedo are higher near the poles. Some of this increased reflectivity is absorbed by cloud coverage, which is particularly high during these same seasons. April and May have the highest levels of solar radiation in the world's snow-covered areas and therefore, have the greatest effect on the global radiative balance.

Thermal Diffusivity

Thermal diffusivity refers to the rate at which heat can transfer through a particular object. It is determined by dividing density and heat capacity. In layman terms, it is also known as how "cold to the touch" a particular object may be. Heat travels significantly slower through ice and snow than through air. This means that snow and ice help insulate the ground and water below from heat transfer. Heat is even slower when the snow and ice cover reaches between 30 and 40 centimeters. This works to keep everything under the snow and ice slightly warmer during winter months and slightly cooler during summer months. Thermal diffusivity also plays an important role in the climate.

Latent Heat

Latent heat refers to energy that is released or stored in constant-temperature conditions. For example, the latent heat required to melt ice is relatively high. In other words, in the cryosphere, latent heat is the energy required to change the state of water (from gas to liquid to solid). The heating and cooling of snow and ice coverage contributes to weather changes across the globe. As water evaporates from the earth's surface, it becomes moisture in the atmosphere. For example, the summer monsoon season in Eurasia is thought to be caused by the cooling characteristics of snow and moist soil during the spring.

Types of Cryosphere

The cryosphere refers to all frozen areas around the world. This includes: snow, sea ice, freshwater ice, frozen ground, and glaciers.

Snow

Snow makes up the second largest area of the cryosphere, covering over 18 million square miles. Most of this area is in the Northern Hemisphere and ranges from 17.9 million square miles in the winter to just 1.46 million square miles during the summer. North American snow coverage has remained nearly the same over the majority of this century despite increased springtime temperature. However, this is not the case in Eurasia, where snow coverage has decreased.

Melting mountain snow coverage contributes most of the water to streams and groundwater around the world. This helps explain why mountains make up approximately 40% of global protected areas. Researchers expect global climate change to affect precipitation levels and the amount and timing of snowmelt. This will, in turn, affect worldwide water management procedures.

Sea Ice

Large portions of the ocean near the north and south poles are covered in ice. In the Southern Hemisphere, sea ice covers between 6.56 million square miles and 7.7 million square miles in September. In February, this number can drop to as low as 1.15 million square miles. The seasonal variation is not so stark in the Northern Hemisphere. Sea ice in the Arctic region has been on a steady decrease of approximately 2.7% each 10 years from 1978 to 1995. From 1978 to 2012, that measurement changes to a 3.8% decrease. The Antarctic region, however, has indicated an increase of around 1.3% every decade.

Freshwater Ice

Freshwater ice can be found in rivers and lakes. Typically, it is a seasonal occurrence and not usually found year-round, like sea ice. Because this ice coverage occurs per season and over a significantly smaller area, its effect on climate is minimal. Records of annual ice coverage and breakup can, however, indicate changes in global climate. This is particularly true of lake ice. River ice breakup is a less reliable source of climate change information because it is largely influenced by both changes to water flow and surrounding temperatures.

Frozen Ground

Frozen ground includes areas with permafrost. In the Northern Hemisphere, frozen ground covers an area of around 20.84 million square miles. Areas of permafrost are not as easily measured, but estimates suggest it covers 20% of the land area in the Northern Hemisphere.

In warmer seasons, the depth of frozen ground has been shown to influence both hydrologic and geomorphic events. The influence of permafrost, however, has yet to be identified. This is because permafrost consists of both ice and soil and rocks at freezing temperatures. The temperature of Alaskan permafrost has increased by 2.4 °C over the last few decades.

Glaciers

Glaciers and ice sheets are considered part of the cryosphere. Both consist of large ice masses that sit on top of land. These ice masses melt, become thinner, and spread wider as they move across land. Approximately 77% of the world's freshwater is found in ice sheets. The water in glaciers and ice sheets may remain frozen for between 100,000 and 1 million years.

Role of Cryosphere in Climate

The cryosphere plays a significant role in determining the Earth's climate as a result of the albedo of the light-coloured ice and snow. The high albedo of the different components of the cryosphere reflects a large majority of the incoming solar radiation, helping to regulate the Earth's temperature by balancing Earth's energy budget. Since icy polar regions are extremely sensitive to any shifts in the climate - especially the currently increasing temperatures - the loss of ice and snow from the cryosphere causes a decrease in the area of white surfaces, leading less energy to be reflected and more to be absorbed. This process warms the Earth even more. The melting of Arctic ice is especially concerning as it leads to a positive feedback cycle.

The loss of ice in the cryosphere as a result of an increase in global surface temperature also has several damaging impacts on human life. The loss of ice in the cryosphere is detrimental as this ice is composed of freshwater - not saltwater - and thus provides people worldwide with drinking water. People living in dry areas near mountains in places such as South America and South East Asia rely on the meltwater from glaciers and ice packs for drinking water. Additionally, many rivers that provide people with water are fed at least partially from melting ice. If the cryosphere is reduced too drastically, there will be significantly less meltwater as ice quantities are reduced and a large number of people will have a reduced supply of drinking water. Similarly, there are places in Asia and even in Europe that direct meltwater to irrigate their crops.

References

- Hydrology-definition-scope-history-and-application, hydrology, water: yourarticlelibrary.com, Retrieved 10 March, 2019

- Hydrology-and-its-branches: researchgate.net, Retrieved 17 May, 2019

- Hydrosphere, science: britannica.com, Retrieved 7 January, 2019

- Hydrosphere, geography: eartheclipse.com, Retrieved 12 July, 2019

- Impact-of-human-activities-on-the-hydrosphere, hydrosphere, science: britannica.com, Retrieved 20 February, 2019

- What-is-the-cryosphere: worldatlas.com, Retrieved 23 June, 2019

- Cryosphere: energyeducation.ca, Retrieved 3 April, 2019

Chapter 2
The Hydrological Cycle and Water Balance

The continuous movement of water under, over and on the surface of the Earth is known as the hydrological cycle or the water cycle. The flow of water into and out of a system is described using an equation called a water balance equation. This chapter closely examines the key concepts of water balance and the hydrological cycle to provide an extensive understanding of the subject.

Hydrological Cycle

Hydrological cycle is the cyclic movement of water containing basic continuous processes like evaporation, precipitation and runoff as Runoff ⇒ Evaporation ⇒ Precipitation ⇒ Runoff.

Water moves into and from the various reservoirs on, over, and under the surface of the Earth, and in the process transforms into its various phases of solid (ice), liquid (water), and gas (vapor), with the total mass of water remaining fairly constant. The physical processes of evaporation, condensation, sublimation, precipitation, transpiration, and runoff are responsible for sustaining the water cycle. Heat energy is also exchanged during the cycle, with this store and release of heat affecting climates worldwide. The water cycle is highly crucial to maintaining life on Earth, as it replenishes the world's freshwater resources and moderates extremes in climate.

Description of the Hydrologic Cycle

Evaporation

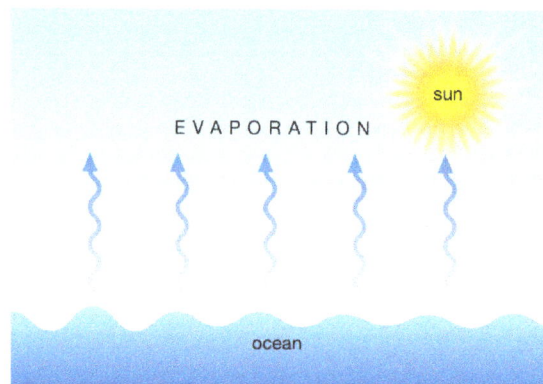

Evaporation occurs when the physical state of water is changed from a liquid state to a gaseous state. A considerable amount of heat, about 600 calories of energy for each gram of water, is exchanged during the change of state. Typically, solar radiation and other factors such as air temperature, vapor pressure, wind, and atmospheric pressure affect the amount of natural evaporation that takes place in any geographic area. Evaporation can occur on raindrops, and on free water surfaces such as seas and lakes. It can even occur from water settled on vegetation, soil, rocks and snow. There is also evaporation caused by human activities. Heated buildings experience evaporation of water settled on its surfaces. Evaporated moisture is lifted into the atmosphere from the ocean, land surfaces, and water bodies as water vapor. Some vapor always exists in the atmosphere.

Condensation

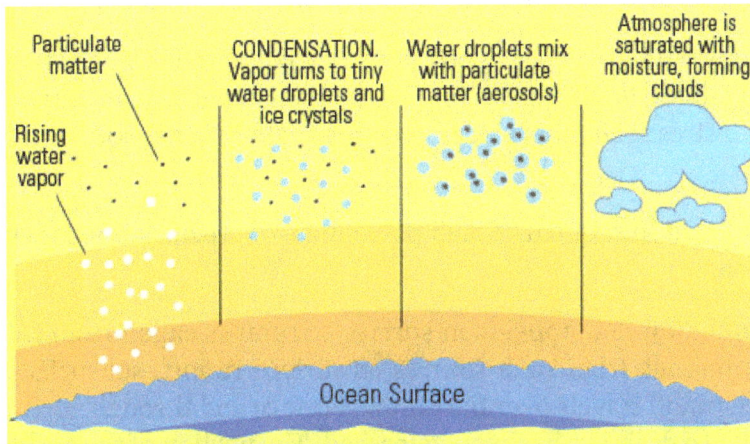

Condensation is the process by which water vapor changes it's physical state from a vapor, most commonly, to a liquid. Water vapor condenses onto small airborne particles to form dew, fog, or clouds. The most active particles that form clouds are sea salts, atmospheric ions caused by lightning,and combustion products containing sulfurous and nitrous acids. Condensation is brought about by cooling of the air or by increasing the amount of vapor in the air to its saturation point. When water vapor condenses back into a liquid state, the same large amount of heat (600 calories of energy per gram) that was needed to make it a vapor is released to the environment.

Precipitation

Precipitation is the process that occurs when any and all forms of water particles fall from the atmosphere and reach the ground. There are two sub-processes that cause clouds to release precipitation, the coalescence process and the ice-crystal process. As water drops reach a critical size, the drop is exposed to gravity and frictional drag. A falling drop leaves a turbulent wake behind which allows smaller drops to fall faster and to be overtaken to join and combine with the lead drop. The other sub-process that can occur is the ice-crystal formation process. It occurs when ice develops in cold clouds or in cloud formations high in the atmosphere where freezing temperatures occur. When nearby water droplets approach the crystals some droplets evaporate and condense on the crystals. The crystals grow to a critical size and drop as snow or ice pellets. Sometimes, as the pellets fall through lower elevation air, they melt and change into raindrops.

Precipitated water may fall into a water body or it may fall onto land. It is then dispersed several ways. The water can adhere to objects on or near the planet surface or it can be carried over and through the land into stream channels, or it may penetrate into the soil, or it may be intercepted by plants.

When rainfall is small and infrequent, a high percentage of precipitation is returned to the atmosphere by evaporation.

The portion of precipitation that appears in surface streams is called runoff. Runoff may consist of component contributions from such sources as surface runoff, subsurface runoff, or ground water runoff. Surface runoff travels over the ground surface and through surface channels to leave a catchment area called a drainage basin or watershed. The portion of the surface runoff that flows over the land surface towards the stream channels is called overland flow. The total runoff confined in the stream channels is called the streamflow.

Interception

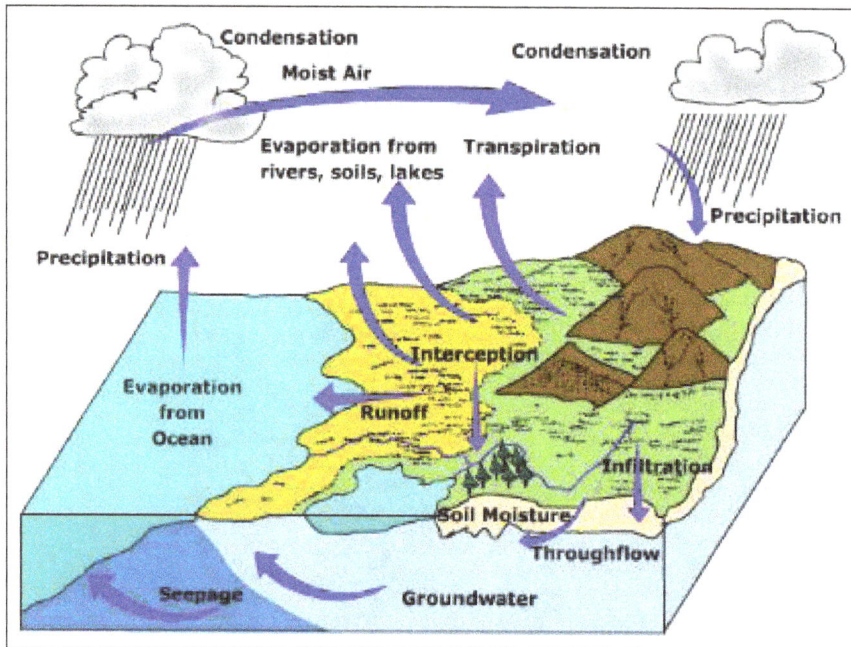

Interception is the process of interrupting the movement of water in the chain of transportation events leading to streams. The interception can take place by vegetal cover or depression storage in puddles and in land formations such as rills and furrows.

When rain first begins, the water striking leaves and other organic materials spreads over the surfaces in a thin layer or it collects at points or edges. When the maximum surface storage capability on the surface of the material is exceeded, the material stores additional water in growing drops along its edges. Eventually the weight of the drops exceeds the surface tension and water falls to the ground. Wind and the impact of rain drops can also release the water from the organic material. The water layer on organic surfaces and the drops of water along the edges are also freely exposed to evaporation.

Additionally, interception of water on the ground surface during freezing and sub-freezing conditions can be substantial. The interception of falling snow and ice on vegetation also occurs. The highest level of interception occurs when it snows on conifer forests and hardwood forests that have not yet lost their leaves.

Infiltration

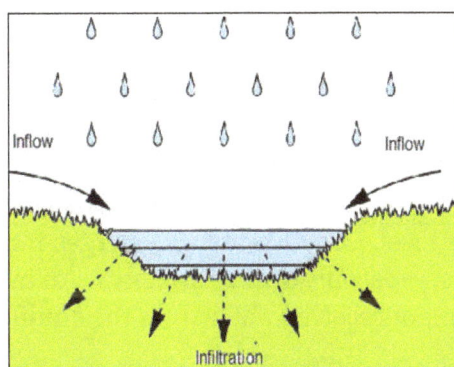

Infiltration is the physical process involving movement of water through the boundary area where the atmosphere interfaces with the soil. The surface phenomenon is governed by soil surface conditions. Water transfer is related to the porosity of the soil and the permeability of the soil profile. Typically, the infiltration rate depends on the puddling of the water at the soil surface by the impact of raindrops, the texture and structure of the soil, the initial soil moisture content, the decreasing water concentration as the water moves deeper into the soil filling of the pores in the soil matrices, changes in the soil composition, and to the swelling of the wetted soils that in turn close cracks in the soil.

Water that is infiltrated and stored in the soil can also become the water that later is evapotranspired or becomes subsurface runoff.

Percolation

Percolation is the movement of water though the soil, and it's layers, by gravity and capillary forces. The prime moving force of groundwater is gravity. Water that is in the zone of aeration where air exists is called vadose water. Water that is in the zone of saturation is called groundwater. For all practical purposes, all groundwater originates as surface water. Once underground, the water is moved by gravity. The boundary that separates the vadose and the saturation zones is called the water table. Usually the direction of water movement is changed from downward and a horizontal component to the movement is added that is based on the geologic boundary conditions.

Geologic formations in the earth's crust serve as natural subterranean reservoirs for storing water. Others can also serve as conduits for the movement of water. Essentially, all groundwater is in motion. Some of it, however, moves extremely slowly. A geologic formation which transmits water from one location to another in sufficient quantity for economic development is called an aquifer. The movement of water is possible because of the voids or pores in the geologic formations. Some formations conduct water back to the ground surface. A spring is a place where the water table reaches the ground surface. Stream channels can be in contact with an unconfined aquifer that approach the ground surface. Water may move from the ground into the stream, or visa versa, depending on the relative water level. Groundwater discharges into a stream forms the base flow of the stream during dry periods, especially during droughts. An influent stream supplies water to an aquifer while and effluent stream receives water from the aquifer.

Transpiration

Transpiration is the biological process that occurs mostly in the day. Water inside of plants is transferred from the plant to the atmosphere as water vapor through numerous individual leave openings. Plants transpire to move nutrients to the upper portion of the plants and to cool the leaves exposed to the sun. Leaves undergoing rapid transpiration can be significantly cooler than the surrounding air. Transpiration is greatly affected by the species of plants that are in the soil and it is strongly affected by the amount of light to which the plants are exposed. Water can be transpired freely by plants until a water deficit develops in the plant and it water-releasing cells (stomata) begin to close. Transpiration then continues at a must slower rate. Only a small portion of the water that plants absorb are retained in the plants.

Vegetation generally retards evaporation from the soil. Vegetation that is shading the soil, reduces the wind velocity. Also, releasing water vapor to the atmosphere reduces the amount of direct evaporation from the soil or from snow or ice cover. The absorption of water into plant roots, along with interception that occurs on plant surfaces offsets the general effects that vegetation has in retarding evaporation from the soil. The forest vegetation tends to have more moisture than the soil beneath the trees.

Runoff

Runoff is flow from a drainage basin or watershed that appears in surface streams. It generally consists of the flow that is unaffected by artificial diversions, storages or other works that society might have on or in a stream channel. The flow is made up partly of precipitation that falls directly on the stream, surface runoff that flows over the land surface and through channels, subsurface runoff that infiltrates the surface soils and moves laterally towards the stream, and groundwater runoff from deep percolation through the soil horizons.

Part of the subsurface flow enters the stream quickly, while the remaining portion may take a longer period before joining the water in the stream. When each of the component flows enters the stream, they form the total runoff. The total runoff in the stream channels is called stream flow and it is generally regarded as direct runoff or base flow.

Storage

There are three basic locations of water storage that occur in the planetary water cycle. Water is stored in the atmosphere; water is stored on the surface of the earth, and water stored in the ground.

Water stored in the atmosphere can be moved relatively quickly from one part of the planet to another part of the planet. The type of storage that occurs on the land surface and under the ground largely depend on the geologic features related to the types of soil and the types of rocks present at the storage locations. Storage occurs as surface storage in oceans, lakes, reservoirs, and glaciers; underground storage occurs in the soil, in aquifers, and in the crevices of rock formations.

The movement of water through the eight other major physical processes of the water cycle can be erratic. On average, water the atmosphere is renewed every 16 days. Soil moisture is replaced about every year. Globally, waters in wetlands are replaced about every 5 years while the residence time of lake water is about 17 years. In areas of low development by society, groundwater renewal can exceed 1,400 years. The uneven distribution and movement of water over time, and the spatial distribution of water in both geographic and geologic areas, can cause extreme phenomena such as floods and droughts to occur.

The Importance of Hydrological Cycle

On the material level, the role of Hydrological cycle is performed in two ways: (1) Physical role and (2) Biological role.

(1) Physical Role: In many ways the physical role is activated:

- The maintenance of balance in exchange of waters between the ground and the sea. Discharge and Recharge of waters - A good amount of waters through rivers and canals or through the evaporation of waters on the ground surface and through carbon synthesis is emitted. Most of waters return to the seas in the form of surface water and the tunnel water and the rest gets evaporated and driven by air returns to the ocean. This process is known as 'Discharge'. If this discharge of water had happened on one side only. The waters of canals and rivers along with the ground water would have dried up. As a result, the plant world would have been extinct and the land would have turned into a desert and the vast expanse of the land would have been inundated with the waters of the sea. But it never happens in reality. The waters of the seas evaporate and rise up and come towards the land driven by wind. Then through condensation, fall by drops to the ground. Then, the canals, tunnels and rivers become full of water and the ground water increases. This is called Recharge. If there were no chance of evaporation, the land would have been flooded gradually. But, as the ratio of discharge and the recharge is almost the same, there is hardly any possibility of inundation or the aridity of the land. Thus, the discharge and the recharge through mutual exchange of waters maintain the balance of the material world.

- Transfer of water in between places in land: Through the amount of evaporation may be less or more in some places on the surface of the land, the water vapour is transferred by the wind. It may happen that where there is more evaporation, the precipitation is less and where there is less evaporation, the precipitation is more. But as there is no hindrance caused by mountains and hills and no arrival of water-vapours from the Seas, it does not rain. Again because of the dry climate of the low river beds on account of their long journey, and want of sufficient rainfall, the deficit of waters is compensated by rainfall, snow-fall and ground water. Thus through the transfer of water and water-vapour, the role of Hydrological Cycle is efficiently performed.

- Role of Hydrological Cycle in landscape evolution: The Hydrological cycle plays the principal role in the changes that take place in the world. Rivers are responsible for the change of the surface of the ground. The source of river water is the Hydrological Cycle. The surface of the land is influenced by ground water and the role of Hydrological Cycle is very important in this as well as in the change of the underground for if there were no recharge, there would have been no ground water.

(2) Biological Role: The biological role in the Hydrological Cycle in preserving the material world is very important. This may happen in many ways:

- Water-supply to the living world: The other name of water is life. Life first appeared in the water (in the seas, 350 crore years ago). For want of water physical world is affected and so life is not possible without water. The ground water (tubewells and wells), the Sweet water of the rivers, and lakes are the source of drinking water.

- Growth of Plant Kingdom: No growth or birth of plants is possible without water. The excessive amount of water creates wood land in the moist areas. Even in dry climate, beside the river-beds or where there is the existence of ground water in the neighbourhood oasis is created. Plants are the refuge of the animal world. Plants inhale the water-vapours through carbon synthesis and add water vapour to the atmosphere and rain takes place.

- Cultivation: By Hydrological process, the soil gets wet – and irrigation works flourish and as a result crops grow in plenty.

- Maintaining Aquatic Eco-system: It is possible to maintain aquatic ecosystem on the land on account of the supply of water in the hydrological cycle. Fishes, Snakes, Frogs, Crocodiles, and other aquatic animals and water – plants like marshes, grasses, water lotuses etc. constitute aquatic eco-system.

- Growth of civilisation and conservation: No animal or plant can live without water, Because of the convenience offered by water in the river – beds, the civilization of mankind centring round cultivation has become possible. In recent times, the system of cultivation has been modernised, industries and technology have developed. As long as the advantage of hydrological cycle is available, human civilization will be progressing leaps and bounds.

The Terrestrial Hydrological Cycle

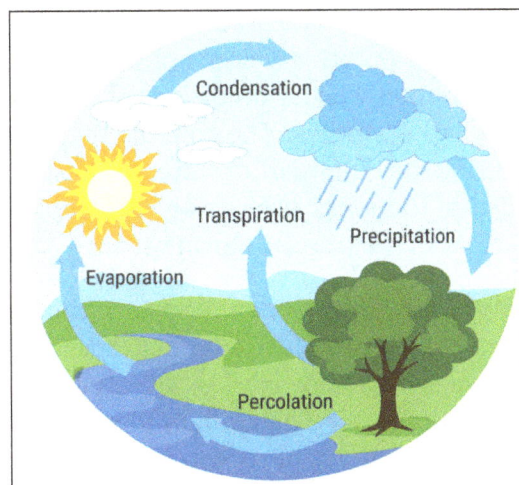

Hydrological Cycle

The key component of the hydrological cycle is generation of river runoff and movement of water in the river networks. The main land area unit where this process occurs is the river watersheds. The sizes of these areas vary from tens of to 6900 square km (the Amazon River catchment area). Within these areas, distinct spatial differences, in topography, geology, vegetation, soil properties, land use, and meteorological conditions may be well-expressed even on small scales. The land surface heterogeneity may be essentially strengthened by human activities that can cause a significant modification of the characteristics of the natural landscapes. Therefore, to describe the hydrological cycle it is important not only to single out the main processes, but also to take into account the relevant topographic, geological, vegetation, and soil parameters that control runoff generation conditions and give an opportunity to represent the land surface heterogeneity.

Precipitation

Precipitation is the principle source of the Earth's water supply and may occur in liquid (rain) and solid (snow) forms. The production of the precipitation results from condensation of small water vapor droplets around available nuclei, or from ice crystal process in the clouds. Water droplets are increased in size by means of collision and coalescence until they attain approximately 2 mm in diameter; under action of gravity they then begin to descend to the Earth's surface forming the rainfall. Ice crystals may also collide and stick to one another, forming snowflakes. These snowflakes can reach the ground in the form of snow or rain, depending on the temperature of the lower atmosphere. For the condensation of water vapor or the creation of ice crystals, it is necessary for the moist air to cool to a sufficient extent and generate lift. Precipitation can be classified into four main types according to the air lifting mechanism: (1) frontal precipitation, where the lifting is due to relative movement of two large air masses; (2) precipitation caused by horizontal convergence; (3) convective precipitation; and (4) orographic precipitation. Each type rarely occurs alone in nature, but some may dominate under certain conditions.

Frontal or cyclonic precipitation occurs at convergence of air masses of various character and at different temperatures. A warm front is formed when warm air rises over the cold air at a relatively gentle slope of 1:100 to 1:400. The precipitation zone extends 300–500 km ahead of the warm front. A cold front is formed when cold air moves under a warm air mass forcing the latter upward. A steeper sloping interface (1:25 to 1:100) is observed. The precipitation zone is limited in this case to about 80 km ahead of the front. The horizontal convergence of air into a low-pressure point results in vertical displacement of air, which may lead to condensation and precipitation. Such meteorological processes commonly occur on or near the tropics as northern and southern components of the trade winds and easterlies. The cold air that commonly prevails over warm oceans in the lower latitudes during the latter part of summer causes tropical storms during which enormous wet air masses pulled in the lower layers rise in the upper atmosphere. The resulting rains fall mostly near the trajectory of the tropical storm center.

Horizontal convergence may also occur as western and eastern sides of two adjacent low-pressure cyclones meet. Frontal and horizontal convergence commonly generates precipitation of moderate intensity. Convective precipitation is caused by local differential heating of air masses, leading to air instabilities and upward movement of air. Instability showers often occur when cold air moves over a warm surface. Air-mass shower is the name of convective rains that are not associated with a pressure system. These showers commonly have relatively low intensity and small areal coverage.

In many regions, a significant part of precipitation is caused by thunderstorms. These convective storms have high intensity and short duration. Thunderstorms develop in three stages. During the first stage, which lasts 10–15 minutes, cumulus cloud formation is observed. Simultaneously, upward air flows at velocities of up to 60–70 km per hour and a significant horizontal inflow of air into convective cells occur. The vertical air movement may reach heights of 7–8 km. The second stage lasts 15–30 minutes and is characterized by strong lifting air movement at velocities to 110–120 km per hour and high rainfall intensity. At heights of 1.5–2.0 km, descending air movement begins. During the dissipating stage, descending air movement predominates until the convective cells disappear.

Orographic cooling occurs when air masses are forced to rise over an obstruction, like a mountain ridge. The result is condensation and rain on the windward side of the mountain, with contrasting dryness on the lee side of the mountain. The amount of precipitation at the orographic cooling is roughly proportional to the wind speed up the slope and to the amount of moisture in the air. Mountains are not so efficient as cyclonic systems in removing the water from a given air mass, because the rising of the moist air caused by mountains is usually less than in cyclonic systems. However, orography is a constant factor in the cause of precipitation at the same place. Regions with orographic effects exhibit relatively high precipitation accumulation, as well as increased frequency of events (for example, some mountain regions of Mediterranean area, the region of the Cascades in the northwestern United States, and some coastal regions of Japan).

Time-spatial distribution of rainfall, especially of storm rainfall, is important for many hydrological events. Storms generally exhibit one or more centers of maximum depth. The difference between the area-averaged depth and the storm-center value increases with increasing area and decreases with increasing total rainfall depth. For storm rainfall in many regions, stable depth–area–duration relations exist. In many cases, it is also possible to construct the dependencies between rainfall frequency, its duration and its average intensity.

Snow Cover and Ice

Permanent snow cover is formed on about 20 percents of the Northern hemisphere and about 15 percent of the Southern hemisphere. A significant part of the land is covered by snow several times during the cold period. Changing the heat balance of the land, the snow cover has a considerable effect on the climate. The presence of snow cover on a drainage basin also greatly influences runoff generation. In many parts of the world, river runoff consists mainly of water yielded by the melting of snow. The snowmelt spring runoff of most large plain rivers of Russia and Canada exceeds half of annual runoff; at the same time, the portion of snowmelt runoff from mountain areas in the arid regions can be significantly larger.

Snowfall over an area is more uniform than rainfall, however; snow accumulation is largely a function of elevation, slope, exposure, and vegetative cover. Snow spatial redistribution is strongly affected by the interaction of wind and topography as well as by interaction of wind and vegetation. Gullies and surface depressions are filled up by snow first of all and can accumulate a considerable portion of the total river basin snow resources (in some parts of Russia, the snow in rills and gullies consists of about 30 percent of total river basin snow resources). In forests, much of the intercepted snow is blown off and accumulates on thesoil surface. The snow retention coefficients (the ratios of snow catch in the surface in question to the accumulation in an otherwise virgin soil) vary from 0.4

for open ice surface and 0.9 for arable land, to 1.2 for hilly district and 3.2 for edges of forests. During blowing and transport of snow significant evaporation(sublimation) may occur (the evaporation losses may reach 40–50 percent of annual snowfall). The snow water equivalent (the depth of water which would result from the melting of the snow) in forest areas is usually 10–40 percent more than in the open areas (in some cases, a general increase of precipitation in the forest is possible). Snow accumulation generally increases with elevation because of the combined effect of the prevailing lower temperatures and the increased frequency of precipitation events caused by orographic effects.

The small-scale variations of snow cover, caused by spatial change of terrain, vegetation, and local meteorological conditions, are superimposed on large-scale variations associated with physiographic and climatic zonality. This leads to very large spatial variability of snow cover characteristics, and they are often considered as random values. The coefficients of spatial variation of the snow water equivalent range from 0.15–0.20 in the forest zone to 0.30–0.60 in the steppe zone. To describe spatial variability of the snow water equivalent one commonly applies the lognormal or gamma statistical distributions.

After snowfall, the snowpack undergoes essential transformation (metamorphosis) caused by compaction, action of the thermal gradients, and change in the crystal structure resulting from interactions of ice, liquid water, and water vapor. Because of migration of water vapor and the freezing together of the small particles of ice, the average ice particle size increases and to the end of winter a snowpack commonly consists of uniform coarse crystals (the process of the formation of coarse snow crystals is called riping). The metamorphosis of snow produced a significant change of density and other physical properties of snow. Snow at the time of fall may have a density as low as 0.01 to as high as 0.15 gcm^{-3}; snowfall in the form of dry snow may vary in density between 0.07 and 0.15 gcm^{-3}; average wind-toughened snow has a density about 0.28–0.30 gcm^{-3}. Ripe snow has a uniform density of 0.4–0.5 gcm^{-3}. The greatest density that can be attained by shifting the snow grains around is about 0.55 gcm^{-3}. Further densification, which can occur under action of deformation, refreezing, and recrystallization, produces a compact, dense material called firn. At a density of between 0.82 and 0.84 gcm^{-3}, the air spaces disappear and the material becomes impermeable to air and water. This material can be defined as ice. The old ice has a density about 0.90 gcm^{-3}; the theoretical density of pure ice is 0.92 gcm^{-3}. Accumulation on land of ice resulting from recrystallization of snow or other forms of precipitation leads to the formation of glaciers. Typical peculiarities of the glacier are the presence of an area where snow or ice accumulates in excess of melting, and another area where the wastage of snow or ice exceeds the accumulation, as well as a slow transfer of mass from the first area to the second. Glaciers exist in a wide variety of forms. They range in size from ice masses occupying tens of square meters to the great continental ice sheets of Antarctica and Greenland. On the Earth's land there are 140 glaciers with areas of more than 1000 square km; at the same time, only on the area of the past Soviet Union are there about 30 000 glaciers of size less 0.1 square km.

A significant amount of ice can accumulate in the ground. If the climate is very cold, a layer of frozen ground may be formed which persists from year to year. The surface layer of this ground (the active layer) normally thaws during the summer and refreezes during the winter, but the ground below remains frozen and impermeable. Such ground is called permafrost and occupies about a quarter of the Earth's land. In areas mantled with peat or a dense mat of living vegetation, the active layer is generally thin and permafrost occurs close to the surface. In areas of bare gravel or exposed bedrock, the active layer may be quite thick. Permafrost is more widely and continuously distributed in lowlands than it is in the mountains in spite of lower temperatures prevailing in the mountains. Lakes, reservoirs, and large ponds produce a warming effect on the ground increasing the depths where lies permafrost.

The water frozen on the land surface and in the ground may form icings which cover considerable areas. In the northeast part of Russia, icings occupy 7–10 percent of area and accumulate 200–300 mm of water.

Being a porous medium, the snowpack has much in common with the soil. In the dry snow, liquid water retains mostly by film tension and capillary forces. The porosity of snow varies from 0.80–0.87 (for new snow) to 0.50–0.70 (for old coarse-grained snow). The liquid water-holding capability of snow (the maximum value of liquid water content beyond which water will drain by gravity action) is about 0.13–0.15. The movement of water through the snowpack begins when the snowpack is saturated by liquid water more than to these values. In the period of snowmelt, a part of the liquid water may refreeze.

The rate of snowpack melt is determined by the incoming heat. The energy budget of the snowpack includes: the net shortwave and long wave radiation; the turbulent exchange of heat in the atmospheric layer above the snow surface (sensible heat); the latent heat consumed in evaporation and sublimation; the heat delivered to snowpack by precipitation; the heat exchange at the land surface; and the change in heat storage including the heat released by freezing of liquid water content. The net shortwave radiation is the most dominant energy component during snowmelt. In the process of metamorphosis and riping the snowpack decreases its reflected capability (albedo) and absorbs the most part of shortwave radiation during snowmelt. The new snow has the albedo 0.75–0.90, and after riping the albedo can reach 0.35–0.40. The empirical dependence can usually be constructed between the albedo and the snow density as a characteristic of snow riping. A close relationship commonly also exists between albedo and the accumulated daily maximum temperature after the last snowfall. The sensible heat is the second important energy budget member. Sometimes, the precipitation heat can be a considerable contribution to positive snowpack energy balance. However, in most cases effects of rainfall on the riping snow and a decrease of albedo are more important.

The most simple and informative index of the snowmelt rate is the air temperature. The relation between these values can be presented as:

$$M = a\left(T_a - T_b\right)$$

where M is the snowmelt in millimeters per day, T_a is the air temperature in degrees Centigrade, T_b is an air temperature below which no melt occurs (it is commonly 0– 2 °C), and a is an empirical coefficient (degree-day factor) which can be interpreted as the snowmelt per day at change of air temperature per degree. The degree-day factor varies depending on climatic and physiographic conditions, but in many cases variations are possible to classify according to the latitude, topography, and vegetation. Because forest cover has a significant effect on many of the variables affecting snow cover energy exchange, there is an essential difference in degree-day factors for forest and open areas. The typical degree-day factors for mid-latitude open areas are usually 4– 5 mm/day °C; for deciduous forest the figure is 3–4 mm/day °C; for dense coniferous forest 1.5–2.0 mm/day °C.

Differences in aspect are also important. At open mountain areas the degree-day factors reach 5–6 mm/day °C. Melt factors in Arctic areas tend to be smaller than those at lower latitudes with similar physiographic conditions, mainly due to lower radiation intensities and relatively little wind during the melt season. Windy areas typically have higher melt factors than areas where calm conditions prevail. In many cases, the degree-day factors increase during the progress of snowmelt as a result of the decrease of the snow albedo, soil warming, and increasing solar radiation. For example, the degree-day factor averaged for Finland is 1.45 mm/day °C at the beginning of the snowmelt period, and 4.75 mm/day °C at the end of the snowmelt period. The maximum values of the degreeday factor reach 80–90 mm/day °C.

The main difference between the melting of snow and ice results from the low albedo of ice. Typical mid-latitude degree-day factors for ice melting are 5–10 mm/day °C. In investigating mass balance of the glaciers, it is more suitable to measure the ablation (that refers to all processes by which solid material is removed from the glacier) instead of the melt. Because evaporation from the glacier surface is small, in temperate climates the values of the glacier melt and of the ablation are close. In some high Arctic regions, appreciable snow and ice are removed by wind erosion. Most ablation occurs on the surface of a glacier. During the ablation season the surface level of a glacier drops not entirely due to ablation, but partly due to compacting, or densification, of the snow layers beneath. Thus, in order to measure ablation, one must measure the thickness and the density of a surface snow layer at each time of measurement.

Interception and Depression Storage

Before reaching the land surface, a part of the precipitation may be intercepted by vegetation and other types of surface cover. A portion of intercepted rainfall evaporates and the other portion may flow down on vegetation stems. Rainfall interception varies with species composition, age, and density of vegetation cover. A dense conifer stand usually intercepts to 25-30 percent of the rainfall at the stem flow of 5–7 percent. The net rainfall interception by hardwood stands is about 15 percent for the period with leaves and about 7 percent for the period without leaves. According to the detailed measurements carried out in the Central Amazonia, the net interception in the tropical rainforest is approximately 10 percent of rainfall. The rainfall interception losses for dense grasses and herbs is as great as for deciduous trees. Interception can be also be significant in large urban areas. The urban landscape includes flat rooftops, potholes, parking lots, cracks, and other rough surfaces that can intercept and hold a significant amount of water.

Interception of snowfall by vegetation may lead to direct sublimation of snow and significant redistribution of snow by wind. The interception of falling snow by conifer forests often reaches 30–35 percent. Snow interception in hardwood stands is about 7– 10 percent.

To calculate net rainfall reaching the land surface through the canopy, the Rutter model of interception is often used. In this model the canopy is considered to have a surface storage of capacity S_c, which is filled by rainfall P, and emptied by evaporation and drainage Q. This capacity may be interpreted as the minimum depth of water required to wet all canopy surfaces. When the depth of water C on the canopy equals or exceeds S_c, the evaporation from the canopy is assumed to occur at the rate E_p. When C is less than S_c, the rate is assumed to be $E_p C S_c^{-1}$. The rate of change of storage is then calculated as:

$$\frac{\partial C}{\partial t} = Q - ke^{b(C-S_c)}$$

Where,

$$Q = a\left(P - E_p C / S_c\right) \text{ when } C < S_c$$
$$Q = a\left(P - E_p\right) \text{ when } C > S_c$$

k, a, and b are empirical constants which depend on the type of vegetation; t is time.

The value of storage of capacity S_c can change over time, depending on seasonal growth of the vegetation cover.

The water reaching the land surface begins to fill up first of allthe depressions of land surface, and simultaneously moves vertically down under action of gravitational and soil suction forces. The sizes and depths of these depressions vary within a large range depending on relief, slope, vegetation, human activity (especially land use and land treatment). The portion of precipitation trapped in the surface depressions can be crudely evaluated as the runoff losses minus the infiltration during the largest rainfall. On the moderate and gentle uncultivated slopes, the initial runoff losses on the filling up of the surface depressions are, on the average, 1.0–1.5 cm; however, some kinds of land treatment can lead to almost full cease of the runoff. In some regions (for example, the tundra), the surface depressions and closed areas can occupy a significant part of river basins. Assuming the portion of area with a depth of depressions exponentially decreases with growth of this depth, we can express the total volume of water stored in depressions as,

$$S = Sm\left[1 - \exp\left(-R / S_m\right)\right]$$

where R is the accumulated excess of rainfall over the infiltration and S_m is the maximum depression storage capacity of the drainage basin, which depends on the physiographic conditions and land use.

Infiltration of Water into the Soil and Vertical Movement of Soil Moisture

Infiltration is the flow of water through the soil surface. The rate and volume of the infiltration depend on the conditions on the soil surface, soil properties (texture, structure, and chemical peculiarities), and soil moisture content.

At the bare soil, a surface crust may develop under action of raindrops. The impact of raindrops breaks down soil crumbs and aggregates, and the particles of silt and clay penetrate previously existing pores, clogging them, and greatly reducing infiltration. The vegetation protects the soil from rainfall action and increases entrance permeability resulting from root activity and increment of the organic content. Root systems perforate the soil, keeping it unconsolidated and porous. The organic matter promotes a crumb structure and improves permeability magnitude. During short high-rate rainstorms, most of the rain quickly travels through macropores to the lower layers of the soil, and only a small fraction of the rain is absorbed by the soil matrix. During low rate storms, a greater fraction of the rain is adsorbed by soil matrix and soil swells reducing the width of the macropores. The swelling soil after drying may also form a surface crust.

Soil texture is determined by the size distribution of individual particles in the soil (the percentages of clay, silt, sand, and coarse fragments more than 2 mm). Soil structure depends on morphological properties of soil particles, and clay, silt, and sand types. It is characterized by bulk density, pore-sized distribution, and construction of vertical profile of soil. The pore sizes and pore-size distribution are greatly affected by the content of soil organic matter, which determines both the sizes of soil aggregates and their stability in water soil texture and structure are closely related to soil porosity and capillary suction forces. Natural cracks, worm holes, or tillage marks create soil macroporosity. The increase of soil porosity leads to the increase of soil permeability, but also to the decrease of capillary suction. Chemical properties of the soil affect the integrity and stability of the soil aggregates, processes of colloidal swelling of the soil, and the suction pressure of the soil matrix.

Water may exist in the soil as liquid water, vapor, or ice. A part of liquid water (hygroscopic and capillary water) is held by molecular forces of the soil matrix. Hygroscopic water exists in the thin films around soil particles at negative (suction) pressures of 31 to 10 000 bars, and may freeze at temperatures below 0 °C. Capillary water is held at a negative pressure of 0.33 to 31 bars, filling gaps between the particles. As the soil moisture increases, the gravitational forces become strong enough to counteract the negative pressures (this occurs at pressures between 0 and 0.33 bars). The maximum amount of water which soil can hold against gravity is called field capacity. Water in excess of field capacity percolates down the soil column, ultimately reaching the soil layer with a small permeability where this movement stops. The plant root system can extract the water if the negative soil moisture pressures are less 15 bars (the content of soil moisture at this pressure is called the permanent wilting point).

Soil water content has a significant effect on the physical and chemical characteristics of the soil and conditions of water flow in soil pores. Soils with an appreciable amount of silt or clay are subject during wetting to the disintegration of the crumbs or aggregates, which in their dry state may provide relatively large pores. These soils also normally contain more or less colloidal material, which in most cases swells appreciably when wet. The pores of sands are relatively stable.

When the all pores of the soil are saturated by water, the water flux depends mainly on gravitational forces. The dynamics of flow in the saturated soil can be described by Darcy's law. This law states that the velocity of flow through a porous medium is directly proportional to the gradient of the piezometric head h = pρg + z, where p is the pressure head, ρ is the density of water, g is gravitational acceleration, and z is the elevation of the point under consideration from an arbitrary datum. If q is the vertical velocity of soil moisture movement, then Darcy's law can be expressed as:

$$q = -K \frac{dh}{dz}$$

where K is the coefficient which is called the hydraulic conductivity saturated soil (or the coefficient of filtration).

The dynamics of flow in the unsaturated soil for isotermical conditions at some additional unessential assumptions can be described by the relation, analogous to Darcy's law:

$$q = -K(\theta) \frac{dh}{dz}$$

where, the pressure head h = $\psi(\theta)$ + z includes the capillary suction ψ (θ) that is as well as the proportionality coefficient K(θ) (the hydraulic conductivity of unsaturated soil) a function of soil moisture, θ is the ratio of the volume occupied by liquid water to the total volume of soil pores (volumetric soil moisture content). Substitution of the last relation into the mass-conservation equation results in a dependence that is called the Richards equation:

$$\frac{\partial \theta}{\partial t} = \frac{\partial}{\partial z}\left[D(\theta)\frac{\partial \theta}{\partial z} \right] - \frac{\partial K}{\partial z} - r(z,t)$$

where $D(\theta) = K(\theta)\,\partial \psi /\partial\theta$ is called the diffusivity of soil moisture, and r is the uptake of water by plant roots.

A lot of different empirical relationships which relate the hydraulic conductivity of unsaturated soil, the diffusivity of soil moisture to the volumetric soil moisture content, and the commonly measured soil moisture constants (porosity, hydraulic conductivity of saturated soil, field capacity, permanent wilting point, and others) has been developed. The uptake of water by plant roots is a complicated biological process, however, which in many cases can be represented as an empirical function of difference of soil and root capillary–osmotic suction, the hydraulic conductivity of unsaturated soil, and the root density.

The moisture movement in unsaturated soil often shows a clear hysteresis effect: the relationships between the soil matrix suction and the soil moisture, as well as between the hydraulic conductivity and the soil moisture, are not the same during wetting and drying events. Capillary forces speed up the filling of small pores during wetting, but delay their emptying during drying. Moreover, the tortuosity of channels, where the water fluxes occur, essentially depends on the previous history of soil moisture conditions.

Freezing of the soil decreases its porosity and capillary forces. If the soil is frozen while saturated, it may become completely impermeable. The hydraulic conductivity of dry soil commonly changes insignificantly after freezing; however, in cold periods a considerable variation of the soil moisture may occur under action of the temperature gradients. There is often a significant flux of soil moisture from the unfrozen zone to the front of freezing. This flux may lead to swelling upper layers of soil and to decreasing soil permeability. In the snowmelt period, the hydraulic conductivity of the dry frozen soil may also decrease as a result of freezing the melt water inside of the soil matrix, where the melt water mixes with the overcooled hygroscopic and capillary water.

The infiltration of melt water into frozen soil can stop because of formation of an impermeable layer at a soil depth. In order to compute heat and moisture transfer in frozen and thawed soil, the following system of equations can be used:

$$\frac{\partial \theta}{\partial t} = \frac{\rho_i}{\rho_w}\frac{\partial I}{\partial t} = \frac{\partial}{\partial z}\left(K\frac{\partial \psi}{\partial z} - K \right)$$

$$C_{ef}\frac{\partial T}{\partial t} = \frac{\partial}{\partial z}\left(\lambda\frac{\partial T}{\partial z} \right) + \rho_w C_w\left(K\frac{\partial \psi}{\partial z} - K \right)\frac{\partial T}{\partial z} + \rho_i L\frac{\partial I}{\partial t}$$

Where

$$C_{ef} = \rho_w C_w \theta + \rho_i C_i I + \rho_g C_g (1 - P_s)$$

θ and I are the volumetric contents of liquid water and ice in the soil, respectively; T is the temperature of the soil, Ps is the soil porosity, K is the soil hydraulic conductivity, L is the latent heat of fusion of ice, and λ is the soil thermal conductivity; ρ denotes density, C is heat capacity, w, I, and g are indices of water, ice, and soil matrix. To take into account the change in soil capillary pressure and the hydraulic conductivity of frozen soil, different empirical formulae can be used.

Depending on the soil moisture conditions and soil–rock properties, the subsurface space where the infiltrated water is stored may be divided vertically into two zones based on the relative proportion of pore space that is occupied by water: an unsaturated zone, or aeration zone, in which the pores contain gases (chiefly air and water vapor); and a saturated zone in which all soil pores are filled with water. During recharge periods, water moves under the force of gravity downward through the unsaturated zone to the saturated zone. The upper limit of the saturated zone (the water table) is the depth at which the water is at atmospheric pressure. The unsaturated zone is often divided into the soil water zone, extending downward from the land surface as far as plant roots penetrate; the capillary fringe, where water rises by capillary forces above the saturated layers; and the intermediate zone, where downward percolation presumably occurs, at least intermittently, toward the saturated zone. The depth of the soil water zone is variable and dependent on soil type and vegetation; the capillary fringe may be from practically zero in coarse material to tenths of meters for fine clays; the intermediate zone may be hundreds of meters thick or be completely absent. Water in the capillary fringe exists at pressures less than atmospheric pressure. All pores may be saturated near the base of this capillary fringe, and the number of pore spaces that are filled with water decreases in an upward direction. In areas of shallow water table, the capillary fringe may extend upward to the root zone or plants and even to the land surface, thus permitting discharge of water by evaporation.

Evapotranspiration

The evaporation of the water into the atmosphere begins when the temperature of the evaporating surface is such that some molecules of the liquid water have attained enough kinetic energy to eject from the water or the land, and to penetrate into the air forming water vapor. Some of these molecules may return from the air and condense, but the number of escaping molecules will be larger than the returning ones until the number of molecules in the air reaches a value which is the maximum possible amount for a given temperature of the air and the water vapor becomes saturated. If more molecules enter the surface than leave it, condensation occurs. The motion of the molecules escaping from the evaporating surface and returning from the air produces a difference of pressure which is determined by the rate of evaporation. The losses of kinetic energy needed for transfer of liquid water into vapor lowers the temperature of evaporating surface (the latent heat of vaporization of water at 0 °C is 596 calg^{-1}). To support evaporation, a steady income of energy has to occur. An energy balance for a given area, particularly for a defined water body can be written as:

$$Q_e = Q_s - Q_r + Q_a - Q_{ar} - Q_{bs} + Q_v - Q_h - Q_c$$

where Q_e is the energy used in evaporation (latent heat), Q_s is the incident solar radiation, Q_r is the reflected solar radiation, Q_a is the incoming longwave atmospheric radiation, Q_{ar} is the reflected longwave radiation, Q_{bs} is the longwave radiation emitted by evaporating body because of its temperature, Q_v is the net energy advected by moving water, Q_h is the heat removed from the system into the air as sensible heat, and Q_c is the change in system energy. The main source of such an inflow of energy during evaporation of water into the atmosphere is solar radiation and the heat of the atmosphere. Evaporation can be estimated by direct or indirect measurement. Direct methods are mostly dominated by point sampling or integrated measurements over small areas, mostly with evaporation pans or lysimeters (vessels or containers placed below the land surface to intercept and collect water moving downward through soil). Indirectly, evaporation can be measured by performing a water balance of a given area.

The rate of evaporation from the water surface into the atmosphere depends on the difference between the pressure of saturated vapor at the temperature of the water and the vapor pressure of the air above the water surface. The last value is determined by the content of the water vapor in the air and the air temperature, which depend in turn on the atmospheric circulation and the turbulent transport in the atmospheric boundary layer. Experimental research has shown that the rate of evaporation from water bodies is not related to the size of the areas of these bodies, if these areas are less than approximately 20–30 square km. The rate from the larger areas slowly decreases with the growth of water body areas, reaching the values of the evaporation rate from the seas and oceans. During the warm period of the year, the rate of evaporation from the water bodies does not necessarily depend on the water body depth if it is more 2–3 meters (there is an insignificant growth of the evaporation rate to these depths). During the cooling of the surface water in the fall and the heating the water body in the spring, the role of the water body depth in evaporation increases. In deep water, the relatively low temperature of the surface water in the spring (less than 4 °C, the temperature of maximum water density) will decrease evaporation, while the higher temperature in the fall leads to an increase in evaporation.

The rate of evaporation from water surface E_w can be calculated as,

$$E_w = C_e V_r \rho_a \left(q_s^* - qa \right)$$

where C_e is a bulk coefficient for water vapor, V_r is the mean wind speed at some reference level z_r above the water surface, ρ_a is the density of the air, q_a is the mean specific humanity at a reference level z_a, and q_s^* is the saturation specific humidity at the mean temperature at the water surface. The value of C_e can, in principle, be calculated on the basis of the characteristics of the atmospheric boundary layer, but given the unavailability of data this coefficient is possible to find more exactly by fitting.

Because the water surface temperature is often unknown, to calculate the rate of evaporation the Penman combined mass-transfer and energy balance method is frequently applied. This requires information only on wind speed, temperature, humidity at one level, and the available energy flux at the water surface. In this method,

$$E_w = \frac{\Delta}{L_e \left(\Delta + \gamma \right)} Q_{ne} + \frac{\gamma}{\Delta + \gamma} C_e V_r \rho_a \left(q_a^* - q_a \right)$$

where $\Delta = dq^* / dT$ at the temperature of the air T_a, q_a^* is the air saturation specific humidity at the same temperature, $\gamma = c_p / L_e$ is the psychrometric constant, c_p is the specific heat of air at constant pressure, L_e is the latent heat of vaporization, and Q_{ne} is the net radiation minus the heat flux to the ground.

The rate of evaporation from the land is determined not only by meteorological conditions but also by the amount and rate of water supply to the evaporating surface. Further, water molecules have to overcome greater resistance to escape from a surface of soil or plant than from a free-water surface. Evaporation of water by plants (this process is called by transpiration) occurs mainly through intercellular openings in the leaves (stomata), which open to allow in the carbon dioxide necessary for photosynthesis and respiration.

To reach the surface of the soil and plant cover, soil moisture must move from the lower depths to the surface. If evaporation from the land is to be a continuous process, movement will have to take place through considerable distances in unsaturated soils. The movement of water upward to the evaporating surface in the soil occurs under action of molecular forces of the soil matrix. In the case of enough dry soil the movement of water vapour under action of the soil temperature gradients may also play an essential role. The wet soil often maintains a practically constant rate of evaporation at a certain range of moisture content, until allow moisture content (approximately the permanent wilting point) is reached. According to some experimental investigations, evaporation from the soil surface continues as long as the upper surface layer—about 10 cm for clays and about 20 cm for sands—remains moist.

The movement of the water to the stomata is caused by stomatal capillary suction and osmotic pressure resulting from the difference in moisture between sap at the roots and surrounding soil. The volume of water evaporation by plants is much larger than the volume of consumption in the formation of vegetative material, and water coming from the roots reaches the stomata almost entirely. Transpiration is a complicated process, depending on both biological and environmental factors. The most important biological factors are type, stage, and growth of the plants, leaf and root structure, and density and behavior of stomata. If the water supply to the leaves is greater than the evaporative capacity of the atmosphere, transpiration is at its climate-controlled potential rate. A plant may help the transpiration process by root development, change soil moisture gradients, and regulate stomatal openings.

In many cases, vegetation cover may also affect the temperature and the moisture content of the atmospheric boundary layer. The link between transpiration, photosynthesis, and the exchange of carbon dioxide makes the transpiration an important factor in long-term interactions of vegetation and climate. Many different attempts at f modeling the processes of energy and water transfer in the soil–vegetation–atmosphere system have shown that the construction of such models require more than 50 parameters reflecting the local soil, vegetation, and atmosphere conditions. To determine most these parameters, it is necessary to carry out special measurements that are only available at the present time for small areas.

Analogously to equation $Q_e = Q_s - Q_r + Q_a - Q_{ar} - Q_{bs} + Q_v - Q_h - Q_c$, the transpiration rate in the simplified form can be presented as,

$$E_T = \rho_a \cdot A_f \left[a_f^* \left(T_f \right) - q_{ac} \right] / r_{af}$$

where A_f is the ratio between the area of leaves and the land area shadowed by leaves (the leaf area index), q_f^* is the saturated special humidity of air in stomatal openings at the temperature T_f, q_{ac} is the special humidity in the canopy air space, and r^{af} is a coefficient which characterizises the aerodynamic and stomatal resistance to water transfer.

The sum total of evaporation and transpiration is usually called evapotranspiration. When the vegetation cover is dense, the transpiration commonly is larger than the evaporation from the soil. According to experimental data collected in the central part of Russia, transpiration contributes 45 percent of evapotranspiration in conifer forest, and 50 percent of evapotranspiration. In deciduous forest, while evaporation from soil is only 30 percent of evapotranspiration in conifer forest, and 35 percent in deciduous forest (about 25 percent of precipitation in conifer forest and 15 percent in deciduous forest is evaporated as a result of interception). In the Amazonian rain forest, on the average, 50 percent of the incoming rainfall is reevaporated, about 25 percent through the interception process and almost the all remainder by transpiration. Transpiration is the predominant cause of losses of soil moisture in arid and semi-arid regions.

It is often necessary to differ the actual evapotranspiration and the potential evapotranspiration (the climatically controlled rate of evapotranspiration that occurs when the supply of water to the land surface and water resource in the root zone are unlimited). The potential evapotranspiration is more easily measured and calculated (approximately, it can be estimated as evaporation from the shallow water surface). To calculate the rate of actual evapotranspiration ET_a, it is possible to use the relation.

$$ET_a = \beta ET_p$$

where ETp is a potential evapotranspiration rate and β is a reduction or moisture availability factor. In many cases, it is possible to take,

$$\beta = 1 \qquad \text{for } w > w_0$$
$$\beta = w / w_0 \qquad \text{for } w < w_0$$

where w and w_0 are the soil moisture contents in the 1 m soil layer.

The most known model used for estimation of actual evapotranspiration is the Penman– Monteith equation:

$$ET_a = \frac{\Delta}{L_e\left[\Delta + \gamma\left(1 + r_c / r_a\right)\right]}Q_{ne} + \frac{\rho_a \gamma\left(q_a^* - q_a\right)}{r_a\left[\Delta + \gamma\left(1 + r_c / r_a\right)\right]}$$

where r_a is the aerodynamic resistance to water vapor transport, and r_c is the stomatal resistance to water transport in dry conditions. Coefficient r_a is determined as functions of the roughness of land and vegetation cover as well as the characteristics of the atmospheric boundary layer. Coefficient r_c varies as a function of soil moisture as well as vegetation type and equals to zero for a wet canopy.

Evaporation from snow requires a three-phase change of state from a solid to liquid to gas (this process is called sublimation). The latent heat of sublimation is much higher than the latent heat of melting (80cal g^{-1}), so that the latter is a preferred process. In order for evaporation to occur the saturated vapour pressures at the snowmelt temperatures must be low enough and the air above the snow surface must be sufficiently dry. If there is a favorable vapour pressure gradient, evaporation of snow may occur even at absence of heat income. The necessary heat may be taken from the snow itself by cooling it. On the average, the evaporation rate from snow cover is approximately 0.3 mm per day before snowmelt, and 0.4–0.5 mm per day after the beginning of snowmelt. In spite of the fact that the daily rates of evaporation from snow cover seldom reach more than 1–2 mm, the accumulated evaporation during sunny and dry winter–spring periods may significantly decrease the snow resources before the snow melt (for example, in the forest zone of Russia the evaporation from snow cover leads to decreases of the snow water equivalent before the spring snowmelt by 15–20 percent).

The role of condensation on land surface in the hydrological cycle has been investigated insufficiently. However, it is known that in some mountain regions condensation of liquid water from fogs in forest may increase the annual precipitation up 7–10 percent. In arid areas, night condensation of air moisture may essentially increase moisture content in the upper layer of the soil.

Groundwater and Groundwater Flow

Water in the saturated zone of soil–rock systems is commonly called groundwater, and it represents the largest liquid water store of the hydrological cycle. Water may penetrate into soil–rock systems in the process of vertical movement from the unsaturated zone; as a result of filtration from river channels, lakes, and reservoirs; and also as a consequence of artificial recharge Groundwater reservoirs in permeable geological formations that can release a considerable amount of water with relative ease are called aquifers.

If, after drilling a fully-penetrating well through a geological formation, the groundwater rises to the piezometric level (which is equal to the elevation above a datum plus the pressure in the aquifer) this formation is called a confined, or artesian, aquifer. An unconfined (phreatic) aquifer has a free water surface. This free water surface may be directly connected to a stream or other surface waters. The water in phreatic aquifers comes from direct rainfall recharge over the aquifer, from connections to surface waters, and from other aquifers. The confining beds separating the aquifers may be completely impermeable (aquifuge), or "leaky" (aquiclude). Whether a rock or soil formation is an aquifer, aquifuge, or aquiclude depends largely on its geologic origins and history.

Confined aquifers recharge through areas where the soil system is exposed to the surface, or through aquicludes. Many confined aquifers contain "fossil waters" deposited in past geologic times. The best aquifers are generally sediment deposits of alluvial or glacial origin. Some sandstone and sedimentary rocks may have very little permeability through pore openings (for example, dolomite and limestone), and their water-bearing capacity and transmission depend mainly on the degree of fracturing resultant from weathering, and the degree of solution of cementing material. The formation of fractures, crevices, or caves in highly-weathered and dissolved limestone (karst processes) often leads to development of underground river systems.

Movement of groundwater occurs under action of hydrostatic head with velocities ranged from several meters per day to only several meters per year. The change of moving water mass in a confined aquifer can only be attributed to changes of porosity caused by compression of the soil–rock matrix. Sub standing Darcy's law for the groundwater fluxes in all cartesian coordinates x, y, and z in the mass balance expression, and taking into account the change of water storage in the pores, we receive:

$$\frac{\partial}{\partial x}\left(K_x\frac{\partial h}{\partial x}\right)+\frac{\partial}{\partial y}\left(K_y\frac{\partial h}{\partial y}\right)+\frac{\partial}{\partial z}\left(K_z\frac{\partial h}{\partial z}\right)=S_0\frac{\partial h}{\partial t}$$

where K_x, K_y, and K_z are saturated hydraulic conductivities in the x, y, and z directions respectively, and So is the specific storativity (the volume of water released from storage per unit volume of aquifer per unit change in pressure head), which has units of inverse distance and is usually taken as a fitting parameter.

Under steady-state conditions, equation $\dfrac{\partial}{\partial x}\left(K_x \dfrac{\partial h}{\partial x}\right)+\dfrac{\partial}{\partial y}\left(K_y \dfrac{\partial h}{\partial y}\right)+\dfrac{\partial}{\partial z}\left(K_z \dfrac{\partial h}{\partial z}\right)=S_0 \dfrac{\partial h}{\partial t}$ becomes:

$$\frac{\partial}{\partial x}\left(K_x \frac{\partial h}{\partial x}\right)+\frac{\partial}{\partial y}\left(K_y \frac{\partial h}{\partial y}\right)+\frac{\partial}{\partial z}\left(K_z \frac{\partial h}{\partial z}\right)=0$$

For an isotropic and homogeneous medium, the hydraulic conductivities can be taken out of the derivatives and divided out; thus equation $\dfrac{\partial}{\partial x}\left(K_x \dfrac{\partial h}{\partial x}\right)+\dfrac{\partial}{\partial y}\left(K_y \dfrac{\partial h}{\partial y}\right)+\dfrac{\partial}{\partial z}\left(K_z \dfrac{\partial h}{\partial z}\right)=0$ is transformed into the well-known Laplace equation:

$$\frac{\partial^2 h}{\partial x^2}+\frac{\partial^2 h}{\partial y^2}+\frac{\partial^2 h}{\partial z^2}=0$$

Description of the flow in an unconfined aquifer is complicated by the presence of a free surface which changes in time and in plane, and may include a recharge of the water from the unsaturated zone If it is possible to assume that there is only horizontal flow, and that the slope of the phreatic surface is small in comparison to total aquifer depth, for this aim the Boussinesq equation is commonly applied:

$$S_0 \frac{\partial h}{\partial t}=\frac{\partial}{\partial x}\left(K_x H \frac{\partial h}{\partial x}\right)+\frac{\partial}{\partial y}\left(K_y H \frac{\partial h}{\partial y}\right)+R_g$$

where $h(t)$ is the phreatic surface level, $H(t)$ is the saturated thickness, and R_g is the instantaneous vertical recharge into the saturated zone.

If the groundwater is located close to the land surface, it can intensively interact with the surface water. The rise of groundwater level to the land surface may lead to a sharp increase of overland and subsurface flow. Groundwater discharges on the river slopes or in the basin depressions form springs and creeks. Where a river channel is in contact with an unconfined aquifer, groundwater may flow from the aquifer into the river channel, or vice versa, depending upon where the water level is lower. During a flood period, groundwater levels may be significantly raised near a channel by inflow from rivers. This process is known as bank storage. The reduction of the maximum discharge during floods caused by bank storage can reach 10–15 percent. After the rise of river stage during flood, a long period of groundwater recession may be observed. In many cases, pumping of groundwater leads to a decrease of surface runoff. If the groundwater located deep enough, recharge of the unconfined aquifer may occur from the river drainage network without hydraulic interaction. Such a type of recharge is often observed in dry regions or on permafrost.

Groundwater discharging into a river system forms the base runoff that is the main sustainable portion of total runoff for many plain rivers. The contribution of groundwater to the total river runoff may vary from a negligible fraction (for instance, for mountainous rivers) to 100 percent (for some karst river basins); however, there is a clearly-expressed physiographic zonality in the distribution of this contribution. For example, in the northern regions of the European part of Russia, where the water table is shallow and river drainage is not well-developed, the portion of the groundwater runoff is 10–30 percent; in the middle part of Russia, where there is shallow groundwater and well-developed river drainage, the portion of the ground runoff reaches 40–50 percent; and in the southern part, where the groundwater is deep, this portion is 15–30 percent. The portion of groundwater which discharges directly into large lakes, seas, and oceans is about 2 percent; however, in some regions there is a significant submarine intrusion of seawater into coastal aquifers.

River Runoff and Generation Mechanisms

River runoff is that part of the precipitation which is collected from a drainage basin or watershed and flows into the river system. From the hydrologic point of view, the runoff from a drainage basin may be considered as a product of the hydrologic cycle and a result of a compound interaction of meteorological and physiographic factors. Physiographic factors can be classified into two main groups: basin factors (size, shape, and topography of drainage area, geology, properties of soils, presence of lakes and swamps, vegetation cover, and land use); and channel factors (slope, hydraulic properties of the channels, channel storage capacity, sediments, and stream bed material). Frequently two basins of nearly the same size may behave entirely differently in runoff phenomena. The essential differences occur between large and small basins. For example, most large basins have significant channel storage effects that smooth the variations of water inflow caused by meteorological factors or change of conditions on the basin area. Small basins are very sensitive both to climatic factors and change in land use.

The variety of runoff generation conditions is reflected in the temporal-spatial change of runoff coefficients (runoff–precipitation ratios). Depending on meteorological and physiographic conditions, these coefficients may vary from 0 to almost 1. In the deserts of the tropical and temperate zones almost all precipitation evaporates. Small runoff coefficients (0.05–0.15) are also typical for the steppe and dry savanna zones. In the zone of hard-leaf forests, the runoff coefficients are of the order of 0.1–0.2. However, in the zone of permanently-humid forests, the runoff coefficients reach 0.40–0.45. High runoff coefficients are characteristic also for the tundra and rainforest zones (0.5–0.6). The runoff coefficients from glaciers are usually close to 0.8–0.9.

Runoff commonly shows a well-expressed seasonal variability. Runoff of a typical river basin in the temperate climate region has one or several periods with a significant rise in runoff discharges (such rises are usually called floods), and one or several periods of low flow. In the humid and tropic climates, seasonal variability is comparatively less; in arid regions, there are ephemeral rivers where the runoff is non-existent during periods without precipitation, although it may appear at different times.

The variability of runoff can be estimated in terms of the day-to-day fluctuation of the river discharges or stages. A graph showing river discharge with respect to time is known as a hydrograph. Hydrographs can be regarded as integral expressions of the physiographic and climatic characteristics that govern the relations between water inflow and runoff of a particular drainage basin. The shape of a hydrograph reflects the difference in runoff components and their paths of movement. A typical, single-peaked, simple flood hydrograph consists of three parts: the approach segment; the rising (or concentration) segment; and the recession (falling or lowering) segment. The lower portion of the recession segment is a groundwater recession (or depletion) curve, which shows the decreasing rate of groundwater inflow. The peak of a rainfall flood hydrograph represents the highest concentration of the runoff from a drainage basin. It occurs usually at a certain time after the rain has ended, and this time depends on the size and the shape of the drainage basin as well as the spatial distribution of the rainfall. The multiple peaks of a hydrograph may occur in any basin as the result of multiple storms developing close to each other. If a hydrograph shows double or triple peaks fairly regularly, this may be due to non-synchronization of the runoff contributions from several tributaries to the main stream. The recession segment represents withdrawal of water from storage after all inflow to the channel has ceased.

According to the area of genesis, river runoff components can be divided into surface runoff, subsurface runoff, and groundwater runoff. The surface runoff is that part of runoff which is produced on the land surface and flows over the land surface and through river drainage system to reach the basin outlet. The part of the surface runoff which does not reach stream channels is called overland flow. The subsurface runoff, also known as interflow, is that part of the precipitation which infiltrates into the soil and moves horizontally through the soil and the ground above the main groundwater level. A part of the subsurface runoff may enter the stream quickly, while the remaining part may take a long time before appearing in the stream channels. The groundwater runoff is that part of the groundwater which discharges in the river drainage system.

The proportions of surface, subsurface, and groundwater components in the total runoff strongly vary in space and time and are defined by the physical mechanisms of river runoff generation. Field research of runoff genesis on experimental and representative river basins can commonly only provide data which is sufficient for discovering the main features of these mechanisms for small plots. At the same river basin, several distinct runoff generation mechanisms may exist. To establish the leading runoff generation mechanisms on large basins, it is usually necessary to use long series of meteorological data and runoff measurements together.

It is easy to establish from a simple analysis of flood hydrographs that the river runoff includes three components which have differences in timing: 1) quick flow, consisting of water which reaches the river channel network promptly after rainfall or snowmelt and has velocities of several centimeters per second; 2) flow, consisting of water which reaches the river channel network at velocities of order 0.1–1.0 centimeters per second and 3) slow flow, where velocities can be several orders less than the velocities of quick flow. It is commonly assumed that quick flow is mainly overland flow and the slow flow is mainly ground flow. Hypotheses on the paths of the second mentioned component of runoff may be very different, but in most cases, taking into account the velocities, it is possible to determine that it is dominantly subsurface flow.

To explain the mechanism of flow generation, the renowned American hydrologist Robert E. Horton assumed that the overland flow was generated on all (or a significant part) of the watershed area as sheet flow, and only when an excess of rainfall (or snowmelt) over infiltration was formed. In the initial period of rain, all water may infiltrate into the soil, but the infiltration rate decreases as a function of time because of increases in the soil moisture content at the soil surface. At some point in time (it is called the ponding time), the infiltration rate drops below the rainfall rate. The accumulated water covers all the drainage area by a thin layer and then begins to flow along the slope to the rills and gullies. Thus, the necessary conditions for the generation of overland flow by the Horton mechanism are: (1) a rainfall rate greater than the hydraulic conductivity of the soil;

and (2) a rainfall duration longer than the required ponding time for a given initial moisture profile. Field research of rainfall runoff generation confirms that such a mechanism is often observed during highly intensive showers on arid and semi-arid watersheds, which lack enough vegetation cover to retain moisture. At a suitable combination of soils and topography, high rainfall rates lead to splash erosion and transport of soil particles by water fluxes. The transported sediments are deposited on the land surface and can significantly decrease soil permeability or form an impermeable crust. Horton overland flow may also occur during snowmelt on the plain watersheds when the permeability of frozen soil is low.

However, the analysis of runoff coefficients and field observations shows that the rainfall Horton overland flow occurs in the temperate climate zone very seldom. Sheet flow is usually observed only on partial areas where the soil profile is saturated before the start of rainfall. In this case, water accumulates on the land surface due to the soil's inability to absorb any more moisture (regardless of the difference between rainfall intensity and infiltration rate), and such a type of overland flow is called saturation overland flow. Typically, saturation overland flow occurs when long-duration rains cover the areas where the initial water table is shallow and it can quickly rise to the land surface, or when an impermeable layer is relatively close to the land surface. Such areas are commonly located in valley bottoms, along streams, and near wetlands, but various subsurface conditions can also cause the formation of saturated zones in topographically-high parts of a basin. The area of saturation depends on the season and it can expand and contract during a storm and may differ from storm to storm. Thus, the source area of saturation overland runoff can significantly vary. Basins generating variable source runoff often display the same type of relationship to rainfall and watershed conditions as are recognized for Hortonian overland flow.

Because depths of sheet overland flow are commonly less than 1–2cm, and ranges of velocity variation in time and space are also small, the velocities of overland flow are a one-to-one function of flow depth, watershed slope, and roughness of land surface. As a result, to describe overland flow for a one-dimensional case the following equations can be used:

$$b_s \frac{\partial h_s}{\partial t} + \frac{\partial q_s}{\partial x} = R_e \, b_s$$

$$q_s = \frac{1}{n_s} i_s^{-1/2} h_s^{5/3} b_s$$

where b_s, h_s, q_s, i_s, and n_s are respectively the depth, discharge, width, slope, and Manning roughness coefficient for overland flow, R_e, is the rainfall excess, t is time, and x is the space coordinate.

Subsurface runoff may generate in the unsaturated zone of the soil above the layers with the temporary low permeability, or in the temporary saturated soil layers. In mountainous regions, subsurface runoff is often observed in the rough soil mantles lying above the ground with small hydraulic conductivity. Subsurface storm runoff may also occur through macropores resulting from animal or vegetation action, and in fractures and joints between soil strata. These paths may be enlarged by erosion and sediment transport, and piping drainage systems may be formed. Depending on their origins and the stability of their walls, pipes may vary in diameter from less than 10 mm to more than 1 m. In the unsaturated zone, pipe networks carry water in turbulent flows at velocities which match those for open channels, sometimes over distances of several hundred meters. Pipes also provide bypass routes for water in the saturated zone, essentially increasing seepage velocity. This mechanism of subsurface runoff generation may occur when the capillary fringe in regions of shallow groundwater (usually near streams) becomes quickly saturated, resulting in water flow into the stream. The subsurface water may rise to the land surface and form overland flow, which provides a mechanism for the rapid discharge of subsurface water to stream channels.

The one-dimensional equations for subsurface flow can be presented as:

$$\left(\theta_{mp} - \theta_s\right) b_s \frac{\partial h_g}{\partial t} + \frac{\partial q_g}{\partial x} = R_g b$$
$$= q_g = K_g i_s h_g b_s$$

where h_g, and q_g are the depth and discharge of subsurface flow, R_g is the recharge of subsurface water, θ_s is the soil capillary porosity, θ_{mp} is the maximum porosity including macropores, and K_g is the coefficient characterizing the horizontal hydraulic conductivity of the soil. The main assumptions here are the following: 1) subsurface flow follows the same slope and has the same width of flow strip as the overland flow; 2) the saturated layer h_g is formed above the base of the subsurface ground layer under consideration; 3) the capillary water (i.e. water at a moisture content less than θ_s) does not take part in the horizontal movement.

In spite of an enormous variety of climatic and physiographic conditions which the runoff may generate, in most cases it is possible to establish the general peculiarities of runoff generation for large physiographic zones and types of landscapes (for example, tundra, forest, steppe, arid and rainforest zones, and urban, agricultural, or forest lands). Special mechanisms of runoff generation are typical for mountain, swamp, permafrost, and glacier watersheds.

Mountainous regions cover more than 20 percent of the land and provide the main source of available water resources in many arid and semi-arid areas. Mountain watersheds commonly have a well-expressed vertical zonality in climatic and physiographic conditions. The complex structure of mountain topography, and its interaction in blocking and uplifting large air masses, results in widespread and intensive precipitation on windward slopes with great seasonal variation. A considerable increase of precipitation with altitude can generally be observed, but the value of this increase varies depending on climatic zones and exposition of mountain ranges. Mountain topography strongly affects the spatial distribution of water and energy, and generates heterogeneity at all scales. Large variations of albedo, soil, and water storage conditions in relation to the surface conditions (rocks, snow, vegetation, altitude, exposure, etc.) cause local variations in the structure of the atmospheric boundary layer and heat fluxes. At mountain heights greater than 1 km, meteorological processes are influenced by the state of stratification of the atmosphere. Runoff of many mountain rivers is of mixed rainfall and snow melt origin. A characteristic feature of mountain rivers is extreme seasonal variation. Most runoff is produced quickly as overland flow, or subsurface flow in shallow rough ground layers. Immediately after rain or snow melt, destructive floods transporting significant amounts of hard material and sediments can occur. In dry periods, subsurface flow often results in increased soil moisture in lower slopes and valleys, giving better-developed vegetation than that on the upper slopes. The ground flow is generally small. The runoff coefficients are high (0.4–0.6) and vary within a narrow range.

Persistent swamps occupy only about 2 percent of the land surface, but in some regions of the world (for example, central parts of South America and the northwest area of the European part of Russia) swamp watersheds contribute a significant portion of runoff. A characteristic feature of swamp watersheds is that the water table is situated closely to the land surface, so that the run-off varies only marginally during the warm period. However, swamp watersheds respond quickly to large rainfalls. Evapotranspiration from swamps is usually considerably higher than from dry neighboring areas and leads to a decrease of annual runoff.

In the permafrost river basins, most hydrological activities occur in the active layer. Because of the relative impermeability of frozen ground, runoff losses are determined by evaporation and water storage in depressions, peat mats, and large-pored soils. The value of free basin storage capacity depends on the antecedent hydrometeorological conditions of the current year, or foregoing years. The year-to-year change in basin water storage can reach 10–15 percent of the annual precipitation. During snowmelt, the main mechanism of runoff generation is overland flow. A part of the melt water can freeze in the snow, in the peat mats, or in the ground during the nightly drop in air temperature, and because of low ground temperatures generally. The water frozen in the surface basin storage and in the active layer of the ground can generate a significant portion of river runoff during the entire warm period. There are river basins where floods have resulted from the melting of ice after cessation of snowmelt. Subsurface flow starts after the beginning of the melt of ice in the ground, increases gradually in line with increases in the depth of thawed ground, and may become the main mechanism of rainfall runoff generation. The groundwater component of river runoff in the permafrost regions is usually small.

A glacier can be considered as a watershed whose characteristics change during the course of a year. In early spring the surface of glacier begins to thaw. Melt water and rain are effective agents of heat transfer and quickly thaw holes in the lower layer. Gradually, the area between the holes also becomes thawed, and the snowpack reaches a uniform temperature at the melting point. The thawed zone gradually moves to higher altitudes. In late spring, the glacier is covered entirely by a thick snowpack. Melt water and rainfall must travel through the snowpack by slow percolation (unsaturated flow), until reaching melt water channels in the solid ice below. In summer, some bare ice is exposed and here there may be surface drainage. In autumn, a dense snow layer covers only part of the glacier and bare ice is exposed over the rest of the glacier. Melt water and liquid precipitation travel very quickly from the surface to the outflow stream. In winter, snow accumu-lates and the surface layer freezes. A small amount of water deep within the glacier slowly drains out during the winter. The lack of a direct relation between precipitation and runoff from a glacier is evident for all seasons except for late summer. The diurnal fluctuation of ice and snow melt usually corresponds to the diurnal fluctuation of discharge from the glacier, and reflects the pecu-liarities of the shape of the glacier.

The River Network and Movement of Water in River Channels

The construction of a river network and river channels is determined by climate, topography, and the geological structure of a river basin. At the same time, river networks and river channels exhibit amazing general regularity and organization. Each river network has a treelike structure with the outlet stream as the main trunk and tributaries that bifurcate into smaller and smaller rivers. It has been established that there are stable empirical laws controlling the construction of the river network. Stream links can be classified as follows: 1) channels that originate at a source are defined to be first-order streams; 2) when two streams of order ω join, a stream of order (ω+ 1) is created; 3) when two streams of different order merge, the channel segment immediately downstream is taken to be the continuation of the higher order stream. These laws can be summarized as:

Law of stream numbers $\dfrac{N_{\omega-1}}{N_\omega} = R_B$

Law of stream lengths $\dfrac{\overline{L_\omega}}{L_{\omega-1}} = R_L$

Law of stream areas $\dfrac{\overline{A_\omega}}{A_{\omega-1}} = R_A$,

where N_ω, L_ω, and A_ω are, respectively, the number of streams of a given order ω, their mean length, and the mean area contributing to streams of this order; R_B is called the bifurcation ratio and normally ranges from 3–5; R_L ranges from 1.5–3.5; R_A varies from 3–6. The river length is related to the river area by a power function where the exponent is about 0.6. The fact that this exponent is not 0.5 (river length is not proportional to the square root of area) results from the fact that river basins become more narrow as they enlarge. Geometrical characteristics of river channels are determined by climatic and topographic conditions; however, the velocity, width, and depth of flow can be presented as power functions of discharge whose exponents change in narrow ranges.

During movement in the river system the runoff may significantly increase due to the lateral inflow, or vary in time due to change of runoff generation on the watershed. As a result, for the most part of the year river have an unsteady flow varying in space and time. The unsteady character of river flow can especially be observed during floods, or at some river reaches where there are river flow control constructions.

To describe the unsteady flow of river channels, the Saint–Venant equations are usually applied. The basic and general assumptions underlying the development and the applicability of these equations are: 1) flow is gradually varied, or the vertical velocities are considered small in comparison with the longitudinal velocities; 2) the velocity distribution along a vertical, in unsteady flow, is the same as in steady flow; 3) the friction resistance in unsteady flow is the same as in steady flow. The Saint–Venant equations can be written in the following form:

$$\frac{\partial A}{\partial t} + V\frac{\partial A}{\partial x} + A\frac{\partial V}{\partial x} = 0$$

$$\frac{\partial V}{\partial t} + V\frac{\partial V}{\partial x} + \frac{g}{A}\frac{\partial(\overline{y}A)}{\partial x} + \frac{Vq}{A} = g(S - S_f)$$

where A is the area of channel cross section, V is the water velocity, q is the lateral inflow per unit channel length, h is the depth of flow, g is acceleration of gravity, and S is the channel slope. The term S_f is the friction slope which can be determined as $S_f = V^2 P/C^2 A$ (this relationship is known as Chesy's law for the turbulent flow), C is Chesy's roughness constant, P is the wetted perimeter of channel. The first equation is the continuity (or mass-conservation) equation, and the second is the momentum equation. The first term in the left-hand side of the second equation $\partial V/\partial t$ corresponds to local inertial acceleration or velocity change. The second term $V(\partial V/\partial x)$ is the inertial acceleration, corresponding to velocity change in space. The third term is momentum change induced by pressure differentials related to flow depth changes. The fourth term is the momentum change caused by the incoming mass of lateral inflow. The terms on the right are the gravity force and the friction force related to channel slope and roughness. The inertia terms and the term representing the influence of lateral inflow are usually negligible in comparison with those of bottom slope and friction for an uncontrolled river regime.

For many rivers with large and moderate slopes where dams are absent, the term related to flow depth changes is also small. In this case $S = S_f$, and there is a one-to-one relation between the river discharge $Q = AV$ and the area of channel cross-section (or river stage). Thus, the unsteady river flow equations can be written as,

$$\frac{\partial A}{\partial t} + \frac{\partial Q}{\partial x} = q$$

$$Q = \alpha A^m$$

where α and m are constant coefficients (the dependencies $\alpha = CS^{0.5}$, m = 3/2 are commonly used). This approximation of unsteady flow description is called the kinematic wave model.

The equation $\frac{\partial A}{\partial t} + \frac{\partial Q}{\partial x} = q$ can be also presented as:

$$Q = \alpha A^m$$

$$\frac{I}{C}\frac{\partial Q}{\partial t} + \frac{\partial Q}{\partial x} = q$$

Where,

$$c = \frac{dQ}{dA}\Big|_{x=xc}$$

is the celerity of movement of the discharge Q in direction x in any xc (this is known as the celerity of kinematic wave). Following from the relation between the river discharge and the area of channel cross-section, the celerity of kinematic wave increases with increase of the discharge, and, as a result, the forward part of the kinematic wave steepens and the back part attenuates. The maximum discharge of kinematic waves does not change. Kinematic wave behavior is very close to the behavior of most natural floods.

In cases where the term related to flow depth changes is relatively large (the slope of river channel bed may be mild, or there may be backwater effects related to flood control or a significant in-flow), the Saint–Venant equations can be simplified to the so called diffusion analogy model:

$$\frac{1}{C}\frac{\partial Q}{\partial t} + \frac{\partial Q}{\partial x} = \frac{D}{C}\left(\frac{\partial^2 Q}{\partial x^2} - \frac{\partial q}{\partial x}\right) + q$$

where C has the same meaning as in the kinematic waves, and D is a coefficient of diffusion coefficient that introduced the attenuation of the flood waves. The diffusion coefficient is derived from the equation,

$$D = \left(B\frac{\partial S}{\partial Q}\right)^{-1}$$

where B is the width of the channel.

Modeling the Hydrological Cycle of a River Basin

By constructing mathematical models of the hydrological cycle it is possible to chrystallize our conceptual understanding of individual processes of water turnover in nature, and obtain means for diagnosis and prediction of these processes. Complex hydrological processes usually occurs over large areas, and the special heterogeneity of these events, together with a deficit of information on the characteristics of their environment, make the development of models of the hydrological cycle one of the most difficult problems of geophysics. The optimal structure for models describing the individual processes of the hydrological cycle depends not only on the required refinement of these processes, but to a significant extent on the availability of data needed in order to determine the parameters (coefficients) of the models. In many cases, information needed for choice of the structure of the model and determining its parameters are absent, and there are only experimental field data received at the input and the output of the hydrological system under consideration. Less frequently, the structure of the model is chosen mainly on the basis of a priori knowledge of the process or system behavior.

For most practical tasks, the structure of hydrological models and their attendant parameters is determined on the basis of both a priori (theoretical) and experimental information. As a result, according to the proportion of a priori and experimental information for model construction, the models of processes of the hydrological cycle can be conditionally divided into three types: 1) black box models where only measurements at the input and the output of the hydrological systems are used; 2) gray box models (in hydrology, the term "conceptual models" is used more often), where measurements at the input and the output of the hydrological systems provide the main information, but conceptual understanding of processes is also applied; 3) physically based models, in which the choice of the structure is based on a priori information on processes of the hydrological cycles (mainly, on fundamental laws of hydrophysics and hydrodynamics), and the parameters are determined mainly on the basis of direct measurements of characteristics of the hydrological systems.

However, because of strong spatial variability of these characteristics, present-day measurement methodologies cannot provide the accuracy that is necessary for most hydrological problems.

There are also always some inconsistencies between real natural processes and their representation in the models. As a result, to improve the accuracy of the model, some parameters have to be fitted. Such fitting (it is also called calibration) is commonly carried out by applying the comparisons of calculated and measured components of the hydrological cycle. Because runoff is the most exactly-measurable component of the hydrological cycle, in most cases runoff hydrographs at the outlets of river basins are used for fitting, and the construction of models of the hydrological cycle is commonly carried out for river basins with the available measurements of runoff. For fitting the parameters, the available measurements of evapotranspiration, soil moisture, and snow characteristics are also often used.

To decrease the number of fitting parameters, the hydrological systems in the models of the first and second types are usually assumed to be lumped (characteristics of systems and their conditions do not change in space), and, as a result, the models of hydrological processes are also frequently classified as either lumped or distributed. Lumped conceptual models, which are mostly applied in hydrological practice, contain aggregated empirical parameters that have a complicated physical interpretation and a large range of variation. In contrast, distributed physically-based models include parameters with clear physical meanings, and if there are no direct measurements of these parameters in a given river basin, they can often be gained from laboratory or field investigations of hydrological processes in similar physiographic conditions. A good correspondence between the model structure and the prototype may be supposed to facilitate fitting the model parameters and increase the predictive capability of the model. In many cases, even the a priori information on the possible range of parameter values can considerably decrease uncertainty in the estimation.

Another important advantage of physically-based models is the opportunity to use simulation to explore different assumptions and physical hypotheses about the particular basin and mechanisms of the hydrological processes. Such sensitivity analyses, coupled with observed data, allow for the "deciphering" of dominant mechanisms, and simplify the choice of model structure. The development of databases that include information with high resolution in space (in the form of digitized maps, digital elevation models, and remote sensing data from satellites, airplanes, ground-based weather radars, and more modern Geographic Information Systems (GIS)) that can help in handling this information, essentially widens the opportunities of construction of distributed physically-based models of the hydrological cycle. The use of general physical laws and meanings of parameters in physically-based models also simplifies the coupling of models of the hydrological cycle with models of meteorological and other geophysical phenomena.

Human Influence on the Hydrological Cycle

Human activities that change the land cover of river basins and are aimed at regulating the water fluxes in nature can considerably change the hydrological cycle of the separate river basins, and even of large regions. A striking example of such change is the present-day situation in the Aral Sea basin, where intensive irrigation has resulted in almost full cessation of the water inflow from the Syr-Darya River and the Amu-Darya River, as well as the drastic drop in the Aral Sea level. Other well-documented examples include the increased drought risks in the Mediterranean and the Sahel, following removal of vegetation by forest clearing and overexploitation respectively. There are also some indications that the considerable changes in scale and frequency of flooding in the Ganges basin may be explained by deforestation in the local mountainous region.

Due to human activities, the natural hydrological cycle of most river basins is becoming more and more transformed and regimented. The main stream flow regulation methods are construction of dams, levees, barrages, and dikes, which provide water accumulation, decreasing flood flow, and increasing low flow. The major effects of reservoir construction on the hydrological cycle (excepting runoff control) are an increase of evaporation and a rise of groundwater table. In dry regions, evaporation losses from the reservoir water surface may be so large that they seriously compromise any potential gains. At the same time, in the conditions of moderate climate, the reservoir losses on evaporation are relatively small. For instance, evaporation from the reservoirs in the Volga River basin (where there are about 300 reservoirs the storage capacity over 1 000 000 cubic meters) constitute less than 3–5 percent of the Volga River runoff. The rise in groundwater level along the reservoir periphery and in surrounding areas changes the runoff generation mechanism on these areas. Gradual change of the river flow regime can occur as a consequence of decreasing the river's ability to transport sediments, especially in upper parts and in reservoirs. Reduction of sediment input at the dam site reduces the river channel slope and the bed sheer stress, resulting in dropping flow velocities and the development of river meandering.

The impact of irrigation on the hydrological cycle is especially revealing in the arid regions, but it is also considerable in regions with moderate climate where irrigation is of supplementary character. Diversion of water for irrigation purposes from surface or groundwater resources modifies the natural hydrological processes. It is common for runoff and evaporation from irrigated areas to increase significantly. Irrigation in river basins where there is no additional method of supply often leads to runoff reduction in the outlet site. In many dry regions, a considerable rise in the groundwater table can occur because of water filtration from reservoirs, leakage from water distributing systems, and faulty irrigation technology. Such a rise may cause waterlogging of plants and development of soil salinization.

To remove excess water from waterlogged soils, drainage is applied in many regions of the world. The primary effect of drainage is the lowering of the groundwater table and the extension of the layer with unsaturated soil. As a result, evapotranspiration may considerably drop (in some cases, by more than 50 percent). The improvement of hydraulic conditions due to drainage increases flow velocities. In the first years after construction of a drainage system, the annual runoff can increase by 20–30 percent. Especially large runoff rises can be observed during the low runoff months in winter and summer. Acceleration of flow also leads to a significant increase in flood peaks. After 10–15 years the impact of drainage on runoff decreases.

Because the quality of groundwater is mostly far better than that of surface water, and its temperature is relatively constant, large volumes of groundwater are extracted for domestic and industrial use in different regions of the world. If groundwater is extracted from confined aquifers below impermeable layers, the groundwater table is not, or is only slightly, affected. However, at some river basins the groundwater table often drops steeply, and this may reduce the surface runoff and the lower level of the small rivers. In many coastal areas, the extraction of groundwater leads to the seawater intrusion.

Together with direct change of the hydrological regime of the river basins by means of stream channel control, irrigation, drainage, and groundwater abstractions, changing the land use of river basins can exert significant influence on the hydrological cycle. Consequences of land use change may be revealed gradually, and be masked by climate variations, but an essential transformation of hydrological regime can occur. The most significant distortions of the hydrological cycle are observed in urbanized areas. The replacement of natural land cover by the urban impermeable surface causes great reductions in infiltration and evapotranspiration. The rainfall runoff from urbanized areas is mainly generated as overland flow and reaches the river drainage system very quickly. Accordingly, the rainfall flood volumes may increase by several times, and the peaks of the hydrographs may increase by 10–15 times. At the same time, snow transport may result in a decrease of snowmelt runoff. Due to reduction of infiltration and groundwater abstractions for urban water supply, falls in the groundwater table are also observed in urban neighborhoods.

The effects of agricultural and forestry practices on the hydrological cycle are less apparent, and depend, to a significant extent, on the physiographic and climatic conditions. It is evident that ploughing, especially contour ploughing usually breaks up overland flow and increases infiltration. Some special types of ploughing may increase the depression and detention storage on gentle slopes from about 8–10mm (in the natural conditions) to 30–40mm. Tillage and the activity of plant root systems modify the structure of the upper soil layer and change not only the vertical permeability, but also the water retention capacity. Extension of vegetation cover and the leaves area increases the interception of precipitation and evapotranspiration. Control of overland flow by dense permanent grasses on steeper slopes can reduce storm runoff from small watersheds by 20–25 percent. However, the relative influence of all these changes on annual and flood runoff is determined by the river basin and climate characteristics. In dry regions (for example, in the steppe zone), the 15–20 percent change of the annual runoff caused by agricultural practice has been fixed, and in different years these changes reached 30–40 percent. At the same time, in the wet regions, especially in the forest and northern forest steppe zone, the impact of agricultural practices on runoff may be neglected.

The main clearly-expressed effects of deforestation on the hydrological cycle of a river basin are the increases in transpiration and interception of precipitation, which in turn result in a decrease of the volume of total runoff. Deforestation reduces infiltration and improves the conditions for overland flow. As a consequence, flood runoff and peak discharges may significantly increase. At the same time, the higher infiltration of forest soils increases the opportunity for recharge groundwater, and the flow of small rivers tends to be more sustained, especially in the case of the generation of snowmelt runoff, when forests further sustain flow by delaying the snowmelt. A rise in the groundwater table and an increase of ground runoff may also raise the low flow of medium- and large-sized rivers. Such effects often result in the conclusion that forests increase runoff. However, careful observations on representative and experimental basins do not commonly confirm such conclusions. For example, results of fifteen individual watershed-scale experiments, involving various rates of forest cutting, carried out during 50 years at the Coweeta Hydrologic Laboratory in Southern Appalachia, indicated that deforestation increased and afforestation decreased annual and monthly runoff, but the magnitude of the responses was highly variable. It was established that streamflow response to forest cutting was inversely proportional to solar energy input (as an evapotranspiration index). The alteration of the monthly runoff is in close agreement with changes in evapotranspiration. At the lowest flow, the monthly runoff was about 100 percent greater from the clearcut forest than the uncut forest. Clearcutting has little effect on flow during winter and early spring.

Long-term observations have also shown the strong dependence of runoff volume on the type of vegetal cover. Conversion of hardwood to pine reduced the annual runoff by 25 cm and produced significant reductions of monthly runoff. At the same time, forest cutting has led to a considerable increase in flood peaks. Similar results have been also received on the basis of analyzing data obtained in other physiographic conditions. Research carried out in the forest zone of the European part of Russia has shown that the influence of the forest on evapotranspiration and runoff significantly depends on the age of the forest. Cutting of old forest may not alter evaporation, and the increasing accumulation of snow may even lead to some growth of spring runoff. In many regions, deforestation has resulted in a significant increase in disastrous floods and has also caused severe soil erosion.

Water Balance

The water balance is an accounting of the inputs and outputs of water. The water balance of a place, whether it is an agricultural field, watershed, or continent, can be determined by calculating the input, output, and storage changes of water at the Earth's surface.

The major input of water is from precipitation and output is evapotranspiration. The geographer C. W. Thornthwaite pioneered the water balance approach to water resource analysis. He and his team used the water-balance methodology to assess water needs for irrigation and other water-related issues.

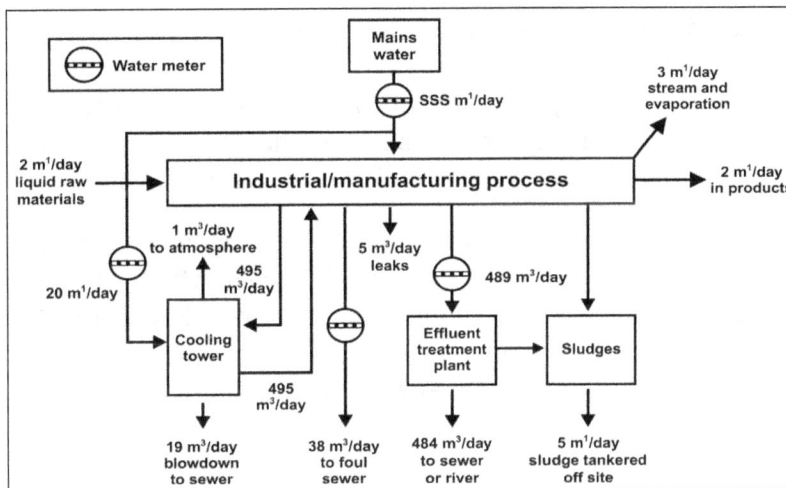

A water balance diagram will help your organisation to understand water use and may help you to reduce water costs. The diagram will include known water use help to identify leaks, overuse and areas where efficiency improvements could be made. This could include the installation of meters (sub-metering) at significant points in the system.

Components of Water Balance

To understand water-balance concept, we need to start with its various components:

Precipitation

Precipitation (also known as one of the classes of hydrometeors, which are atmospheric water phenomena) is any product of the condensation of atmospheric water vapour that is pulled down by gravity and deposited on the Earth's surface. The main forms of precipitation include rain, snow, ice pellets, and graupel. It occurs when the atmosphere, a large gaseous solution, becomes saturated with water vapour and the water condenses, falling out of solution (i.e., precipitates).

Two processes, possibly acting together, can lead to air becoming saturated – cooling the air or adding water vapour to the air. Virga is precipitation that begins falling to the earth but evaporates before reaching the surface; it is one of the ways air can become saturated. Precipitation forms via collision with other rain drops or ice crystals within a cloud.

Moisture overriding associated with weather fronts is an overall major method of precipitation production. If enough moisture and upward motion is present, precipitation falls from convective clouds such as cumulonimbus and can organize into narrow rain-bands. Where relatively warm water bodies are present, for example due to water evaporation from lakes, lake-effect snowfall becomes a concern downwind of the warm lakes within the cold cyclonic flow around the backside of extra-tropical cyclones. Lake-effect snowfall can be locally heavy.

Thunder-snow is possible within a cyclone's comma head and within lake effect precipitation bands. In mountainous areas, heavy precipitation is possible where upslope flow is maximized within windward sides of the terrain at elevation. On the leeward side of mountains, desert climates can exist due to the dry air caused by compressional heating. The movement of the monsoon trough, or inter-tropical convergence zone, brings rainy seasons to savannah climes.

Rain drops range in size from oblate, pancake-like shapes for larger drops, to small spheres for smaller drops. Precipitation that reaches the surface of the earth can occur in many different forms, including rain, freezing rain, drizzle, ice needles, snow, ice pellets or sleet, graupel and hail. Hail is formed within cumulonimbus clouds when strong updrafts of air cause the stones to cycle back and forth through the cloud, causing the hailstone to form in layers until it becomes heavy enough to fall from the cloud.

Unlike raindrops, snowflakes grow in a variety of different shapes and patterns, determined by the temperature and humidity characteristics of the air the snowflake moves through on its way to the ground. While snow and ice pellets require temperatures close to the ground to be near or below freezing, hail can occur during much warmer temperature regimes due to the process of its formation. Precipitation may occur on other celestial bodies, e.g., when it gets cold, Mars has precipitation which most likely takes the form of ice needles, rather than rain or snow.

Actual Evapotranspiration

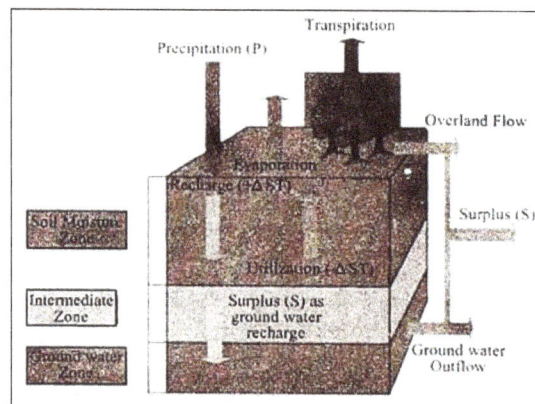

The soil water balance

Evaporation is the phase change from a liquid to a gas releasing water from a wet surface into the air above. Similarly, transpiration is represents a phase change when water is released into the air by plants. Evapotranspiration is the combined transfer of water into the air by evaporation and transpiration. Actual evapotranspiration is the amount of water delivered to the air from these two processes. Actual evapotranspiration is an output of water that is dependent on moisture availability, temperature and humidity.

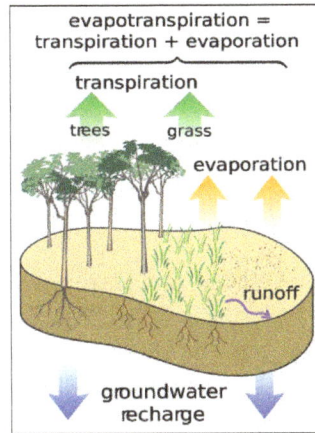

Process of evapotranspiration

Think of actual evapotranspiration as "water use", that is, water that is actually evaporating and transpiring given the environmental conditions of a place. Actual evapotranspiration increases as temperature increases, as long as there is water to evaporate and for plants to transpire. The amount of evapotranspiration also depends on how much water is available, which depends on the field capacity of soils. In other words, if there is no water, no evaporation or transpiration can occur.

Potential Evapotranspiration

The environmental conditions at a place create a demand for water. Especially in the case for plants, as energy input increases, so does the demand for water to maintain life processes. If this demand is not met, serious consequences can occur. If the demand for water far exceeds that which is actual present, dry soil moisture conditions prevail. Natural ecosystems have adapted to the demands placed on water.

Potential evapotranspiration is the amount of water that would be evaporated under an optimal set of conditions, among which is an unlimited supply of water. Think of potential evapotranspiration of "water need". In other words, it would be the water needed for evaporation and transpiration given the local environmental conditions. One of the most important factors that determine water demand is solar radiation.

As energy input increases the demand for water, especially from plants increases. Regardless if there is, or isn't, any water in the soil, a plant still demands water. If it doesn't have access to water, the plant will likely wither and die.

Soil Moisture Storage

Soil moisture storage refers to the amount of water held in the soil at any particular time. The amount of water in the soil depends on soil properties like soil texture and organic matter content.

The maximum amount of water the soil can hold is called the field capacity. Fine grain soils have larger field capacities than coarse grain (sandy) soils. Thus, more water is available for actual evapotranspiration from fine soils than coarse soils. The upper limit of soil moisture storage is the field capacity the lower limit is 0 when the soil has dried out.

Change in Soil Moisture Storage (ÄST)

The change in soil moisture storage is the amount of water that is being added to or removed from what is stored. The change in soil moisture storage falls between 0 and the field capacity.

Deficit

A soil moisture deficit occurs when the demand for water exceeds that which is actually available. In other words, deficits occur when potential evapotranspiration exceeds actual evapotranspiration (PE > AE). Recalling that PE is water demand and AE is actual water use (which depends on how much water is really available), if we demand more than we have available we will experience a deficit. But, deficits only occur when the soil is completely dried out. That is, soil moisture storage (ST) must be 0. By knowing the amount of deficit, one can determine how much water is needed from irrigation sources.

Surplus

Surplus water occurs when P exceeds PE and the soil is at its field capacity (saturated). That is, we have more water than we actually need to use given the environmental conditions at a place. The surplus water cannot be added to the soil because the soil is at its field capacity so it runs off the surface. Surplus runoff often ends up in nearby streams causing stream discharge to increase. The knowledge of surplus runoff can help forecast potential flooding of nearby streams.

Surface Runoff

Surface runoff is the water flow that occurs when soil is infiltrated to full capacity and excess water from rain, snowmelt, or other sources flows over the land. This is a major component of the hydrologic cycle. Runoff that occurs on surfaces before reaching a channel is also called a nonpoint source. If a nonpoint source contains man-made contaminants, the runoff is called nonpoint source pollution.

A land area which produces runoff that drains to a common point is called a watershed. When runoff flows along the ground, it can pick up soil contaminants such as petroleum, pesticides (in particular herbicides and insecticides), or fertilizers that become discharge or non-point source pollution.

Surface runoff can be generated either by rain fall or by the melting of snow, ice, or glaciers. Snow and glacier melt occur only in areas cold enough for these to form permanently. Typically snowmelt will peak in the spring and glacier melt in the summer, leading to pronounced flow maxima in rivers affected by them. The determining factor of the rate of melting of snow or glaciers is both air temperature and the duration of sunlight. In high mountain regions, streams frequently rise on sunny days and fall on cloudy ones for this reason.

Computing a Soil – Moisture Budget

The best way to understand how the water balance works is to actually calculate a soil water budget. We'll use Rockford, Illinois which is located in the humid continental climate of northern Illinois. Rockford lies on the northern edge of the prairie and mixes with deciduous forest. This vegetation has been nearly completely replaced with agriculture. The knowledge of soil moisture status is important to the agricultural economy of this region that produces mostly corn and soy beans.

To work through the budget, we'll take each month (column) one at a time. It's important to work column by column as we're assessing the moisture status in a given month and one month's value may be determined by what happened in the previous month.

Table: Water Budget- Rockford, IL Field Capacity = 90 mm.

	J	F	M	A	M	J	J	A	S	O	N	D	Year
P	50	49	66	78	100	106	88	84	86	73	56	45	881
PE	0	0	5	40	84	123	145	126	85	44	8	0	531
P-PE	50	49	61	38	16	-17	-57	-42	1	29	48	45	
Δ ST	0	0	0	0	0	-17	57	16	1	29	48	12	
ST	90	90	90	90	90	73	16	0	1	30	78	90	
AE	0	0	5	40	84	123	145	100	85	44	8	0	634
D	0	0	0	0	0	0	0	26	0	0	0	0	26
S	50	49	61	38	16	0	0	0	0	0	0	33	258

Soil Moisture Surplus

During December, Rockford is deep in the grip of winter. Potential evapotranspiration has dropped to zero as plants have gone into a dormant period thus reducing their need for water and cold temperatures inhibit evaporation. Notice that P-PE is equal to 45 but not all is placed into storage. Why? At the end of November the soil is within 12 mm of being at its field capacity. Therefore, only 12 millimeters of the 45 available is put in the soil and the remainder runs off as surplus (S = 33).

Total: Soil Moisture Surplus – Rockford, IL Field Capacity = 90 mm.

	J	F	M	A	M	J	J	A	S	O	N	D	Year
P	50	49	66	78	100	106	88	84	86	73	56	45	881
PE	0	0	5	40	84	123	145	126	85	44	8	0	531
P-PE	50	49	61	38	16	-17	-57	-42	1	29	48	45	
AST	0	0	0	0	0	-17	-57	-16	1	29	48	12	
ST	90	90	90	90	90	73	16	0	1	30	78	90	
AE	0	0	5	40	84	123	145	100	85	44	8	0	634
D	0	0	0	0	0	0	0	26	0	0	0	0	26
S	50	49	61	38	16	0	0	0	0	0	0	33	258

Given that the soil has reached its field capacity in December, any excess water that falls on the surface in January will likely generate surplus runoff. According to the water budget table this is indeed true. Note that P-PE is 50 mm and ÄST is 0 mm. What this indicates is that we cannot change the amount in storage as the soil is at its capacity to hold water. As a result the amount is storage (ST) remains at 90 mm. Being a wet month (P > PE) actual evapotranspiration is equal to potential evapotranspiration. Note that all excess water (P-PE) shows up as surplus (S = 50 mm).

Similar conditions occur for the months of February, March, April, and May. These are all wet months and the soil remains at its field capacity so all excess water becomes surplus. Note too that the values of PE are increasing through these months. This indicates that plants are springing to life and transpiring water. Evaporation is also increasing as insolation and air temperatures are increasing. Notice how the difference between precipitation and potential evapotranspiration decreases through these months.

As the demand on water increases, precipitation is having a harder time satisfying it. As a result, there is a smaller amount of surplus water for the month. Surplus runoff can increase stream discharge to the point where flooding occurs. The flood duration period lasts from December to May (6 months), with the most intense flooding is likely to occur in March when surplus is the highest (61 mm).

Soil Moisture Utilization

Table: Soil Moisture Utilization – Rockford, IL Field Capacity = 90 mm.

	J	F	M	A	M	J	J	A	S	O	N	D	Year
P	50	49	66	78	100	106	88	84	86	73	56	45	881
PE	0	0	5	40	84	123	145	126	85	44	8	0	531
P-PE	50	49	61	38	16	-17	-57	-42	1	29	48	45	
AST	0	0	0	0	0	-17	-57	-16	1	29	48	12	
ST	90	90	90	90	90	73	16	0	1	30	78	90	
AE	0	0	5	40	84	123	145	100	85	44	8	0	634
D	0	0	0	0	0	0	0	26	0	0	0	0	26
S	50	49	61	38	16	0	0	0	0	0	0	33	258

By the time June rolls around, temperatures have increased to the point where evaporation is proceeding quite rapidly and plants are requiring more water to keep them healthy. As potential evapotranspiration is approaching its maximum value during these warmer months, precipitation is falling off. During June P-PE is -17 mm.

What this means is precipitation no longer is able to meet the demands of potential evapotranspiration. In order to meet their needs, plants must extract water that is stored in the soil from the previous months. This is shown in the table by a value of 17 in the cell for ÄST (change in soil storage). Once the 17 m is taken out of storage (ST) it reduces its value to 73.

The month of June is considered a dry month (P < PE) so AE is equal to precipitation plus the absolute value of ÄST (P + |ÄST|). When we complete this calculation (106 mm + 17 mm = 123 mm) we see that AE is equal to PE. What this means is precipitation and what was extracted from storage was able to meet the needs demanded by potential evapotranspiration.

There is no surplus in June as the soil moisture storage has dropped below its field capacity. There is still no deficit as water remains in storage. The calculations for July is similar to June, just different values. By the time July ends, water held in storage is down to a mere 16 mm.

Soil Moisture Deficit

Table: Soil Moisture – Rockford, IL Field Capacity = 90mm.

	J	F	M	A	M	J	J	A	S	O	N	D	Year
P	50	49	66	78	100	106	88	84	86	73	56	45	881
PE	0	0	5	40	84	123	145	126	85	44	8	0	531
P-PE	50	49	61	38	16	-17	-57	-42	1	29	48	45	
AST	0	0	0	0	0	-17	-57	-16	1	29	48	12	
ST	90	90	90	90	90	73	16	0	1	30	78	90	
AE	0	0	5	40	84	123	145	100	85	44	8	0	634
D	0	0	0	0	0	0	0	26	0	0	0	0	26
S	50	49	61	38	16	0	0	0	0	0	0	33	258

August, like June and July, is a dry month. Potential evapotranspiration still exceeds precipitation and the difference is a – 42 mm. Up until this month there has been enough water from precipitation and what is in storage to meet the demands of potential evapotranspiration. So, of the 42 mm of water we would need (P – PE) to extract from the soil.

In so doing, the amount in storage (ST) falls to zero and the soil is dried out. What happens to the remaining 26 mm of the original P-PE of 42? The unmet need for water shows up as soil moisture deficit. In other words, we have not been able to meet our need for water from both precipitation and what we can extract from storage. AE is therefore equal to 100 mm (84 mm of precipitation plus 16 mm of ÄST).

So what is a farmer to do if their crops cannot obtain needed water from precipitation or soil moisture storage they irrigate. Irrigation water usually is pumped from groundwater supplies held in aquifers deep below the surface or from nearby streams (if stream flow is sufficient to provide needed water). The amount of irrigation water required is the amount of the soil moisture deficit.

Soil Moisture Seasons

Four soil moisture seasons can be defined by the soil moisture conditions.

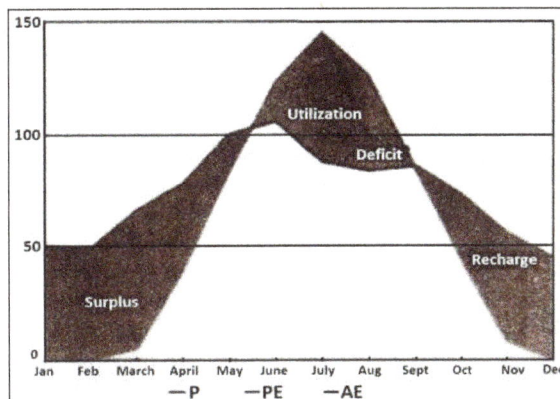

Soil Moisture Seasons for Rockford, Illinois

Recharge

The recharge season is a time when water is added to soil moisture storage (+ΔST). The recharge period occurs when precipitation exceeds potential evapotranspiration but the soil has yet to reach its field capacity.

Surplus

The surplus season occurs when precipitation exceeds potential evapotranspiration and the soil has reached its field capacity. Any additional water applied to the soil runs off. If this water runs off into nearby streams and rivers it could cause flooding. Thus, the intensity (amount) and duration (length of season) of surplus can be used to predict the severity of potential flooding.

Utilization

The utilization season is a time when water is withdrawn from soil moisture storage (−ÄST). The utilization period occurs when potential evapotranspiration exceeds precipitation but soil storage has yet to reach 0 (dry soil).

Deficit

The deficit season occurs when occurs when potential evapotranspiration exceeds precipitation and soil storage has reached 0. This is a time when there is essentially no water for plants. Farmers then tap ground water reserves or water in nearby streams and lakes to irrigate their crops. Thus, the intensity (amount) and duration (length of season) of deficit can be used to predict the need for irrigation water.

Whether a place experiences all four seasons depends on the climate and soil properties. Wet climate and those places with soils having high field capacities are less likely to experience a deficit period. Likewise the duration and intensity of any season will be determined by the climate and soil properties. Given equal amounts of precipitation, coarse textured soils will generate runoff faster than fine textured soils and may experience more intense surplus.

Type of Surface Water Balance

The water balance of the entire mine, a number of components, or a single entity, such as the heap leach pad, may be quantified as part of the water quality and quantity management activities at a mine site.

Reasons for undertaking a facility or site water balance study may include:

(a) Evaluate strategies for optimum use of limited water supplies;

(b) Establish procedures for limiting site discharge and complying with discharge requirements, particularly control of the quality of the water and the quantity of contaminants discharged from the site; and

(c) Limiting or controlling erosion due to flow over exposed surfaces or in channels, swales, and creeks; and

(d) Estimating the demands on water treatment plants, holding ponds, evaporation ponds, or wetlands.

Analytical Approaches

The most common way to build a water balance model of a facility of site is to use that famous stand-by, Excel. The reason is that most water balance models generally involve no more than successive solution for each component of a facility and hence for each facility of the simple equation:

Inflow – Outflow = Change in Storage.

Use of the Mine Water Balance Model

The following are the steps in setting up, refining, and using a water balance model of your mine:

- Model: Have an effective, robust, calibrated and easily updated and adjusted water quality and quantity (volumetric flow) model to understand the complex relationships of the mine for the prediction of water changes. Model all sources of contamination and the inputs as well as outputs. The stakeholders should agree that the model is accurate and appropriate.

- Measure: Have an effective sampling program to keep the water quality and quantity model up to date and continuously evaluate its effectiveness and test its assumptions.

- Calibrate: Have the model checked each week initially (then monthly) with water quality and quantity numbers and monitor discrepancies between reality and model; evaluate and explain discrepancies.

- Contingency Plans: Have a full set of costed contingency plans with established implementation timelines. Complete the engineering for likely long-term options.

- Manage: Understand all possible actions that can be taken to minimise water quality and quantity issues and have them costed to +/− 35% accuracy. Know at what levels what actions need to be taken when pre-specified levels are reached so that management can confidently make decisions which meet its license limits while incurring the least expenditure.

Water Balance Estimation

In the natural environment, water is almost constantly in motion and is able to change state from liquid to a solid or a vapour under appropriate conditions. Conservation of mass requires that, within a specific area over a specific period of time, water inflows are equal to water outflows, plus or minus any change of storage within the area of interest. Put more simply, the water entering an area has to leave the area or be stored within the area. The simplest form of water balance equation is as follows:

$$P = Q + E \pm \Delta S$$

Where, P is precipitation, Q is runoff, E is evaporation and ΔS is the storage in the soil, aquifers or reservoirs.

In water balance analysis, it is often useful to divide water flows into 'green' and 'blue' water. 'Blue' water is the surface and groundwater that is available for irrigation urban and industrial use and environmental flows. 'Green' water is water that has been stored in the soil and that evaporates into the atmosphere. The source of 'green' water is rainfall or 'blue' water has been used for irrigation.

A Water Balance Analysis can be used to:

- Assess the current status and trends in water resource availability in an area over a specific period of time.

- Strengthen water management decision-making, by assessing and improving the validity of visions, scenarios and strategies.

Water balance estimates are often presented as being precise. In fact, there is always uncertainly, arising from inadequate data capture networks, measurement errors and the complex spatial and temporal heterogeneity that characterises hydrological processes. Consequently, uncertainty analysis is an important part of water balance estimation as is quality control of information before used.

When the data sources are imprecise, it is often possible to omit components that do not affect changes. For example, it is possible to omit storage from an annual water balance if year-on-year storage changes (such as reservoirs) are negligible.

Some common problems that occur when water balance estimations are made include:

- Temporal and spatial boundaries are not defined.

- The quality of input data is poor.

- Double counting of water flows when water flows within an area added to water flow exiting area.

- Inappropriate extrapolation of field level information to a larger scale. Many hydrological relationships are scale dependent (e.g. runoff as a proportion of rainfall is almost always higher at smaller spatial and temporal scales).

- Intuition (often based on popular myths) is used rather than good quality information.

- The storage term(s) of the water balance is omitted.

- Political or other pressures result in unreliable estimates that have been manipulated.

Almost everyone is influenced by the water balance estimates because they are often central elements of awareness raising campaigns. Stakeholders directly involved in decision making may require more detailed information about the water available. Given that water balance analysis should be based on a needs assessment, everyone is involved in determining the outputs that are needed. The process of producing water estimates is best undertaken by experienced specialists or by staff who have undergone training and have access to specialist support.

Materials and Resources

Techniques for carrying out water balance estimation range from very simple 'back of the envelope' estimates to highly complex computer-based models. A sound knowledge of hydrological processes of a prerequisite of water balance estimation. It is often advisable for a project or programmes to employ the services of a specialist to produce water balance estimates or, at the very last, to provide specialist advice as and when it is needed. Access to a quality-controlled information base is a good starting point for water balance estimation.

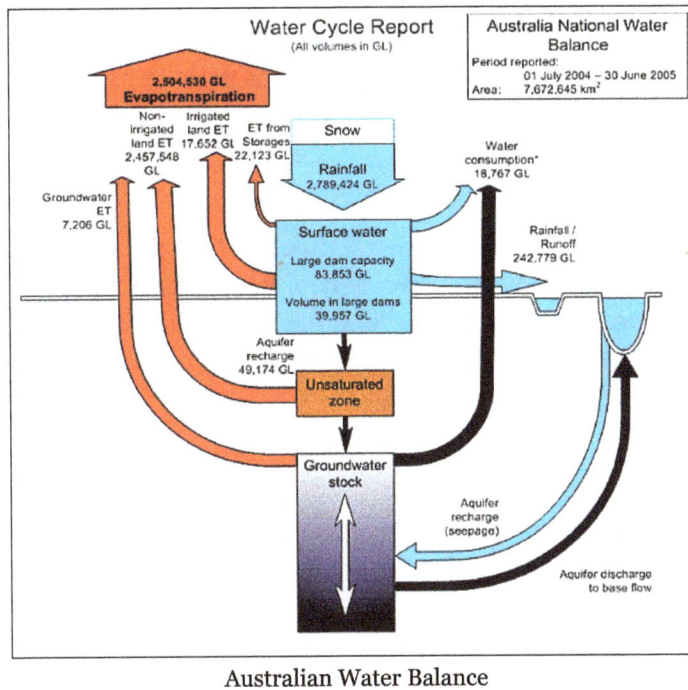

Australian Water Balance

Advantages

- Conducting water balance estimation provides you with a comprehensive understanding of the water flow system and water resources in your area.

Disadvantages

- Published water balance estimates are often incorrect.

- Too often, minimal or no account is taken of uncertainty when estimates are made and presented. Therefore quality assurance and control of the estimates should always be built in to a water balance procedure.

- It is a very complex work process and needs to be done by qualified experts (or at least supported by experts) and this requires considerable time and resources.

References

- What-is-the-water-hydrologic-cycle: worldatlas.com, Retrieved 11 March, 2019

- Hydrology, water-cycle: nwrfc.noaa.gov, Retrieved 1 August, 2019

- 12-Hydrological-Cycle: bhattercollege.ac.in, Retrieved 19 June, 2019

- Kuchment, water: biodiversity.ru, Retrieved 27 January, 2019

- Water-balance-meaning-components-and-types-hydrology-geology, water-balance, hydrology: geographynotes.com, Retrieved 30 April, 2019

- Water-balance-estimation, water-sources, sustainable-water-supply: sswm.info, Retrieved 28 February, 2019

Chapter 3

Groundwater and Surface Hydrology

The branch of hydrology which seeks to study the distribution and movement of groundwater in the rocks and soil is referred to as groundwater hydrology. Surface water hydrology deals with the study of water above the surface of the Earth. This chapter has been carefully written to provide an easy understanding of the varied facets of groundwater and surface hydrology.

Groundwater Hydrology

Groundwater is that portion of the water beneath the surface of the earth that can be collected with wells, tunnels or drainage galleries or that flows naturally to the earth's surface via seeps or springs. Groundwater is a resource of immense magnitude, but of uneven availability and inexhaustibile. The mode of occurrence of groundwater is as varies as the rock types in which they occur and as intricate as the formation of the earth's crust, through geologic time. The occurrence and movement of groundwater is controlled by several factors such as climate, geology, hydrology, topography, ecology, vegetation and soil distribution.

Groundwater constitutes 0.614% of the earth's fresh water resources, compared to the 0.008% in lakes and 0.005% in rivers. Unlike the surface water, the groundwater resource is fairly well distributed. Groundwater sources provide a reasonably constant supply, which is not likely to drier under natural conditions, as surface water sources. Harnessing groundwater is less expensive compared to any surface water irrigation projects. For medium to small-scale consumption like domestic, industrial and agricultural needs, groundwater exploration is much cheaper. In terms of biologic or organic quality groundwater is generally better than the surface water but in terms of chemical quality (dissolved solids), surface water is better than groundwater.

Origin of Groundwater

Atmospheric precipitation is the main source of groundwater. The rainwater might have infiltrated directly into the ground below or it might have fallen somewhere else and recharged through rivers and lakes. Groundwater of atmospheric origin that has been a recent (geological recently) pert of the hydrological cycle is called 'Meteoric Water'. Much older groundwater that is still of atmospheric origin but that has been isolated from the hydrologic cycle for millions of years is called 'Connate Water'. This water typically is groundwater that was already present in the geologic formation when it was formed, such as the water in which the alluvial material was deposited.

Source of Groundwater

Recharge to groundwater body is accomplished by the processes of infiltration and percolation.

Infiltration is the process whereby precipitation and surface water move downward into the soil mantle or rock surface. Percolation is the vertical and lateral movement of water through the various openings in the geologic materials, in response to gravity or change in pressure.

Recharge to groundwater body may result from natural or artificial availability of water on the surface of the ground. Natural sources are rainfall, snow melt, streams and lakes. Artificial sources are leakage from reservoirs, conduits, Septic Tanks etc. Irrigation, Effluent discharged to evaporation or percolation ponds. Main source of groundwater is rainfall.

Divisions of Sub-surface Water

All water that occurs naturally below the earth's surface is called 'sub-surface water' whether it occurs in the saturated or unsaturated zone. The sub-surface part of the hydrologic cycle is classified as soil water, intermediate vadose water, capillary fringe water and phreatic water (groundwater). However there are no sharp boundaries between the various types. The zone within the sub-surface consisting of soil water, vadose water and capillary fringe water is called 'Zone of Aeration'. The zone beneath the zone of aeration is called 'Zone of Saturation'.

Zone of Aeration - The upper zone of the earth's crust between last surface and the water table. The pores, fissures and other interstices of the soil and rocks of this zone contain pellicular and capillary water. Gravitational water commonly percolates into it through this zone. Many voids are occupied by water vapour and other gases.

Zone of Saturation - The part the earth's crust in which all voids are typically filled with water under hydrostatic pressure.

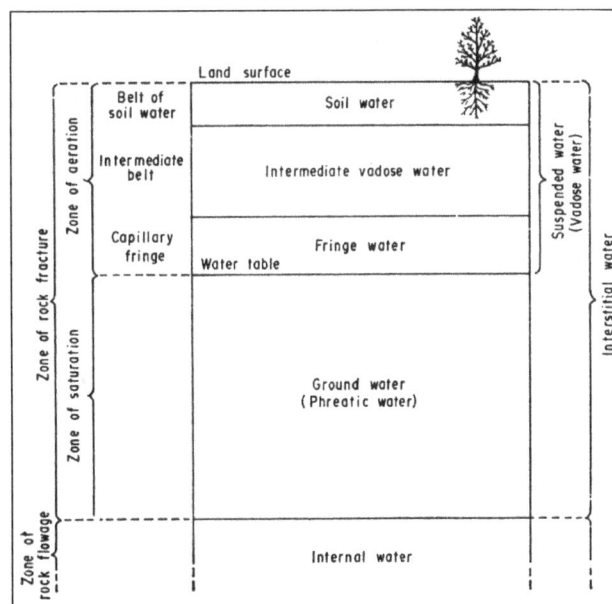

Water Table

The upper limit of the saturated zone in the unconfined aquifer is called 'Water Table'. Water in the saturated zone below the water table is called groundwater. The depth to water table varies from place to place. In general the water is not flat it rises and falls with the ground surface but in a subdued way, so that it is deeper beneath hills and shallower beneath the valleys. Where ever the water table meets the ground surface, the ground will be marshy or there will be a pond, spring or river.

Perched Water Table - If downward movement of infiltrated water is restricted by a deeper layer of low permeability, a temporary unconfined aquifer may form. The water will pond on the restricting layer and form perched groundwater with a 'Perched Water Table'. Perched aquifers may form above a confined or unconfined aquifer. They are usually very limited a really and may not receive sufficient recharge to support significant well production.

Porosity and Permeability

Groundwater can be drawn either from consolidated rocks or unconsolidated loose sediments. The occurance of groundwater in a geological formation and the scope for its development depends on the porosity and permeability of the formation.

Porosity is an inherent property of a rock, soil, etc. containing interstices (pore or open or void space). Porosity is the ratio of the volume of pores to the total volume of the porous material. Porosity depends on the shape, arrangement and the degree of sorting and cementing. Porosity present in a rock may be either primary or secondary. Primary porosity is that which is present in the rock right from its formation. Secondary porosity is that which is present after the rock is formed.

An aquifer of uniform sized, spherical, well-rounded grains will have higher porosity than an aquifer of angular or flat particles. In fractured media, the determinants of porosity are the number of width of fractures per unit volume of rock, the degree to which they are interconnected and the extent to which they are filled with eroded material.

Effective Porosity is the amount of interconnected pore spaces available for fluid transmissions.

Permeability is an intrinsic geometric property of the formation and is related to the ease with which a fluid can move through it. Permeability depends on porosity and interconnection of the pore spaces. The unit of permeability is m/day.

Hydrogeologic Classification of Rocks

The rocks have been classified as aquifers, aquitards, aquicludes and aquifuges based on the porosity and permeability properties:

1. Aquifer: Groundwater bearing formations sufficiently permeable to transmit and yield water in usable quantities. Eg: sand and gravel deposits, fractured rocks, etc.

2. Aquiclude: Formations, which are porous and containing, water but do not allow water to move through them under typical hydraulic gradients. The formations are also called impermeable formations. Eg: clay beds, shales, etc.

3. Aquitard: Formations, which are sufficiently permeable to transmit water vertically to or from the confined aquifer, but not permeable enough to laterally transmit water like an aquifer. (Far less permeable than the aquifer but not impermeable). Eg: Sandy clay.

4. Aquifuge: Formations, which do not hold water or transmit ground water. Eg: Massive hard rocks.

Types of Aquifers

Aquifers have been classified as unconfined, semi-confined and confined aquifers:

1. Unconfined Aquifers: The surface of the groundwater body is at atmospheric pressure. The top of an unconfined aquifer is called 'Water Table'. The groundwater levels are free to rise or fall, as there will not be any impermeable zone on the top. These type of aquifers are also called 'Phreatic Aquifer'.

2. Semi-confined Aquifer: An aquifer bound by one or two Aquitard is called semi-confined or leaky aquifer.

3. Confined Aquifer: Impermeable beds forming boundaries of aquifer or separating various aquifers is called confining layers. An aquifer sandwiched between two confining layers is called 'Confined Aquifer'. Confined aquifers are completely filled with groundwater and they do not have a free water table. The pressure condition in a confined aquifer is characterized by the 'piezometric surface'. If the piezometric surface is above the upper confining layer, then the well is called 'Artesian Well'.

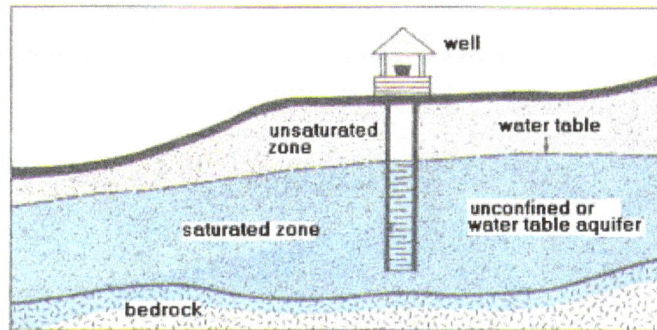

Types of Aquifer

Hydraulic Properties of Aquifer

Other than the porosity and permeability, the other hydraulic properties of aquifers are transmissibility, storage coefficient, specific yield, specific retention, specific capacity and safe yield:

1. Transmissibility: Rate of flow of groundwater through a vertical strip of the aquifer of unit width and depth (or thickness) under a unit hydraulic gradient. i.e T (transmissibility) = k (permeability) X d (saturated thickness). The unit of T is m²/day or cm²/sec or L_2/t.

2. Storage Coefficient: Volume of water yielded per unit horizontal area and per unit drop of water table or piezometric surface. Thus if an unconfined aquifer released 4m³ of water for a water table drop of 2m over a horizontal area of 10m², the storage coefficient is 0.2 or 20 percent.

3. Specific Yield: In unconfined aquifers the storage coefficient can also be called 'Specific Yield', which is the volume of water released from a unit volume of saturated aquifer material drained by a falling water table. Specifc yield is expressed in percentage.

$$\text{Specific Yield} = \frac{\text{Volume of water released by gravity drainage}}{\text{Total volume of saturated aquifer material drained}}$$

4. Specific Retention: The quantity of water retained by the material against the force of gravity. This is expressed as percentage of total volume of the material drained, i.e,

$$\text{Specific Retention} = \frac{\text{Volume of water held against gravity drainage}}{\text{Total volume of material drained}}$$

(Porosity of a rock is naturally equal to the sum of specific yield and specific retention).

5. Specific Capacity: The rates of discharge from a well divide by the drawdown.

6. Safe Yield: The rate at which groundwater can be withdrawn without causing a long-term decline of water table or piezometric surface. Thus, the safe yield is equal to the average replenishment rate of the aquifer.

Darcy's Law

The velocity (V) of the flow of the water through a porous medium is directly proportional to the permeability of the medium(K) and hydraulic head(h) and inversely proportional to the length of flow(L).The law is applicable when the flow is laminar.

V=K h / L,

where

V = Velocity (l/t)

K = Permeability (l/t)

H = Hydraulic head

L = Length of flow (l)

Volume rate of flow, (Q) = VA, where V = Ki,

Q = KiA

Where I (hydraulic gradient) = hl – h²/L or h / L

A = Cross-sectional Area (l²/ t)

Q = Discharge or Volume Rate of Flow (l³/ t).

Groundwater Exploration

The different methods involved in the study of groundwater are Geomorphological, Photo-geological, Geological, Hydrometeorological, Hydrogeochemical and Hydrogeological which includes all these methods to tackle the Groundwater problems.

Mostly the Hydrogeological studies are carried out in basin wise studies. Here all the above said studies will be done with reference to a particular drainage basin and quantitative estimation of the available groundwater in that basin will be calculated.

Remote Sensing Method

This method makes use of Aerial photos and satellite images and this provides a wealth of details of large areas on earth's surface. This method is used for the estimation of surface and sub-surface water over large areas. This method would be extremely useful for rapid hydrogeological mapping of large and inaccessible areas. Areas potential for groundwater development, open water surfaces, springs etc. can be delineated by skillful interpretation.

Geomorphological Method

First with the help of toposheets the drainage basin boundaries are marked following the peak of the hills. This line, which bisects the hilltops, is called as water divide. Next step is to construct contour maps and drainage maps of the basin. By these maps one can delineate the morphologic features such as Hills, Plains and Vallies, whichis of basic importance. Closer the contour steeper the landscape and hence the recharge to groundwater will be very less. If the contour lines are wider apart such areas are characteristic of plains and thus there will be time for rainwater to infiltrate. The same factors can also be detected through the drainage density. If the drainage density is high, it is due to hilly and resistant rocks but if the drainage density is less, it means the land is plain or it has yielded to erosion. Shorter and denser the drainage pattern, the area is not favorable for groundwater development, whereas if the streams are longer and well spread areas are for groundwater development. Even the change in geologic formations, structures and faults can be recognized through the help of drainage patterns. Usually wells drilled at the lowest portion of a valley yield more groundwater when compare to wells drilled at the top of the hills or hill slopes. Similarly the point of intersection of two or more streams holds better chances of yielding good amount of groundwater.

Geological Method

This method is used in detecting different rock types in a given area on the basis of its petrography, stratigraphy and structures. Petrography is the first and foremost important consideration in which the porosity and permeability will be known. The search for groundwater is confined to most promising zones in terms of porosity and permeability (porosity determines the amount of water which can be held in storage and permeability suggests the ease with which water can be extracted). The porosity and permeability will vary a lot amongst different rock types. The areal distribution, thickness and depth of all pertinent aquifers, aquicludes and aquifuges must be determined and maps and cross-sections should be constructed. This will reveal whether there is only one principle water-bearing zone exists, whether it has form of one or several permeable continuous rock layers directly on top of one another or whether there are several Groundwater zones. These predictions can be made with the data of already drilled borehole logs and drawing the same in the form of fence diagram. This diagram even helps in suggesting whether a place is suitable for borewells or dug wells.

The stratigraphy of geologic formations describes the geometrical and age relations between the various lenses, beds and formations in the rocks of sedimentary origin or volcanic origin. Thus stratigraphy is an essential tool in search for water in areas of wide spread sedimentary or volcanic rocks.

The structural studies such as locating joints/fractures in crystalline rocks, faults and folds in sedimentary rocks and many intrusives like sills and dykes, which act as barriers to groundwater movement. The unconformities in the sedimentary rocks are also of particular importance in Hydrogeology. Aquifers are commonly associated with unconformities, either in the weathered or fractured zone immediately below the surface of the buried landscape or in permeable zones in coarse sediments laid down on top of this surface when the system entered a new era of accretion.

Hydrogeochemical Method

This study is also an integral part of the Hydrogeology. It is not only the quality assessment we make but by this method we can also study the predict further changes in chemical composition of groundwater and where these changes may be expected to occur. The water quality may be changes may be expected to occur. The water quality may be changed due to over exploitation or due to industrial effluents or due to over dosage of fertilizers or due to the rocks itself through which the groundwater flows. Hence one should have the natural background in relation to changes in water quality. Some of the chemical parameters could be very successfully be utilized to discuss the areas of recharge, discharge, the residence time of waters within the aquifer, suitability of waters for irrigation, industry or domestic purposes. These factors help the Hydrogeologists to forecast the quality of water to be encountered in an area and to advice the enthusiastic explorer about the suitability of the water even before a well is sunk in any particular area of interest.

Hydrometeorological Method

Hydrometeorological data like precipitation, evapotranspiration and surface runoff are required to determine the water balance of a basin for development and management of its water resources. Detailed study of the rainfall pattern, rate of evapotranspiration and also the amount of runoff makes it possible to access the quantity of groundwater available in an area.

Locating Well Sites

These above mentioned studies help in assessing the potentiality of an area and also reveals the general mode of groundwater occurrence, structural controls and also the geomorphological bearing on groundwater. However, for pinpointing a most favorable location for groundwater exploration, geophysical studies should be carried out.

As the rocks are usually covered by soils, it becomes difficult to assess the potentiality of a place. By using the Geophysical methods, one will be studying the physical properties of the underlying formations like Magnetic susceptibility, Density, Elasticity and Electrical Resistivity. If the contrasts are slight, the measurement may not be precise enough to be useful. Even if the spatial distribution of geologic units is too complex, the results cannot be interpreted geologically. Most of the failures in applying Geophysics to Hydrogeology is from neglecting these factors. So, one has to pick-up the distinct variations in the physical properties of the underlying formations.

Successful application of geophysical methods depends largely on the correct choice of the method or combination of methods; the quality of the instruments; competent fieldwork; correct processing and interpretation of field results and their correlation with geologic data. Without some basis for correlation with observed geologic and borehole data, conclusions from geophysical investigations are likely to remain ambiguous.

The various methods of Geophysical prospecting used for groundwater are the Magnetic, Gravity, Seismic and Electrical. Of all these methods, electrical resistivity prospecting has acquired greatest importance in groundwater investigations and this is mainly due to the large and detectable variation in the resistivity values with the quantity of water in a rock. The resistivity of water bearing rocks largely depends on the amount of water they contain, the chemical composition and temperature, and the distribution of water. The other reasons of preference for electrical resistivity survey are, relative simplicity of instruments and their operation, relatively low operation costs, a depth range commensurate with the depths required in most groundwater problems, suitability to a wide spectrum of subsurface problems and easy methods of interpretation.

Types of Wells

The commonly constructed wells to extract groundwater are (a) Dugwells, (b) Bore wells,(c) Tubewells, (d) Dug-cum-Borewells and (e) Filter point wells. The type of well depends on the quantum of water required, economic considerations and geologic & hydrologic Conditions.

Dugwells

Dugwells are relatively large diameter wells which are dug manually till the water bearing formation is encountered. Mostly these wells are circular in shape, but in some places dugwells will be of square or rectangular shape. The depth, diameter and the shape of these dugwells vary from place to place depending on the local hydrogeological conditions. The diameter of the dugwells range between 1 to 10 m and the depth range between 2 to 30 m depending on the depth to the water table. Usually two types of dugwells are constructed, namely lined wells and unlined wells. In unconsolidated rocks, the walls of the well are lined with concrete rings or brick masonry.

Dug-cum-Borewells

Dugwells are sometimes bored at the bottom to increase its yield by contribution from the deeper horizons. The size of the bore varies from 5-15 cms. In the hard rocks, drilling horizontal, vertical or inclined bores from the bottom of dugwells is a common practice. The success of such drilling depends on the inclination and density of water bearing fractures/joints. The bores should be sufficiently below the pumping water levels to obtain maximum inflow.

Borewells

Borewells are generally drilled in the consolidated formations (Hard rocks). They do not require well assembly to be lowered as practiced in the tubewell construction. Casing pipe is lowered up to the weathered zone and rest of the hole is left uncased. The groundwater is extracted through the fractures in the hard rocks. The depth of the Borewells range between 30 to 300 m.

Tubewells

Tubewells are constructed mostly in unconsolidated and semi-consolidated formations. The construction of tubewells includes drilling of borehole, installing screen and casing pipe & gravel packing. The depth of these wells is governed by location of water bearing zones, yield requirements and costs. The depth of the tubewells range between 20 to 500 m.

Filter Point Wells

Filter point wells are commonly constructed in shallow alluvial formations which are very rich in groundwater. The depth of these wells will be less than 20 metres. The standard size and design of filter points are readily available in the market. Small farmers often go in for these types of abstraction structures to meet out their individual requirements for agriculture. Filter points are used for meeting drinking water requirements too.

Fall in groundwater levels, drying up of wells, ground subsidence, saline-water intrusion (sea-water Intrusion), drying-up of streams / lakes, Induced infiltration / pollution, interference of wells are the major problems faced due to over exploitation of groundwater.

Monitoring of hydrogeologic data such as water level fluctuation, well yield / draft, inspection of wells and chemical and biological quality of groundwater have to be carried out to safeguard the developed groundwater sources and to avoid any environmental problems due to groundwater development.

Surface Water Hydrology

Surface water hydrology includes the study of surface water movement and the distribution of surface water in space and time. Of particular interest is the variability in water quantity and flow within a year and between years. This variability in water supply is largely influenced by climate. Together with geographical characteristics such as topography, soils, and land use, hydrologic variability affects the development and character of surface water systems such as lakes and rivers.

Surface water hydrology is the study of surface water.

Flow variability is also becoming increasingly recognized as an important factor in the health of riverine aquatic ecosystems. Extreme floods are important because many of the processes that shape the river occur during the largest floods, also known as reset events. Hydrologic conditions during extreme low flow periods are also important and can impact species selection. While a river system may remain relatively stable for many years under normal hydrological conditions, the natural variability, including extreme events, is part of the hydrologic regime that creates and maintains a healthy river system.

Glacial Hydrology

Glacial hydrology is a part of surface-water hydrology; some of the runoff from glaciers and snow also involves groundwater hydrology concepts.

Glacier hydrology is the scientific study of the storage and transit of liquid water associated with snow and glacier ice. This includes water present at the surface of, in the interior of, at the basal interface of, and emerging from bodies of snow and ice. Today, glacier ice covers around 11% of earth's surface area, accounts for approximately 70% the planet's freshwater resources, and underpins the water security of over a billion people. For millennia, glacier-fed rivers have been critical for irrigation and agriculture, but have always posed a threat of flooding. Meltwater runoff from glacierized catchments is an important resource for hydroelectric power, accounting for 50% or more of the electricity produced in mountainous countries such New Zealand, Peru, Pakistan, Switzerland, and Norway. Recently, focus has turned to sporadic, catastrophic, and destructive releases of meltwater from glacierized terrain, the mechanisms for and prediction of which remain poorly constrained. Therefore, the hydrology of glaciers is an important, topical subject demanding continued investigation, particularly in the context of the recent trends observed in glacier extents and the earth's atmosphere–climate system.

Glaciers represent a frozen environmental asset: natural reservoirs storing freshwater over a wide range of timescales, from hours to millennia. Long-term freshwater storage, at timescales greater than a year, typically involves glacial ice; intermediate and shorter-term storage more usually involves seasonal snow and liquid water and are best exemplified by the seasonality of glacier runoff. The hydrology and runoff from glaciers is fundamentally controlled by the surface energy balance. For the vast majority of earth's glaciers, outside the tropics, mass is accumulated as snow during the winter, when the energy balance is at a minimum due to low air temperatures and reduced solar radiation receipt. During summer months, increased air temperatures and solar irradiance provide a highly positive surface energy balance, resulting in the melting of snow and glacier ice (termed ablation). Thus, glaciers modulate streamflow by releasing the most runoff during the warmest, driest periods in the year when all other sources of water are minimal. The volume of meltwater runoff that a glacier provides is a function of its surface area and ablation rate. The melt process ensures runoff from glacierized catchments correlates positively with air temperature and exhibits a strong diurnal signal in the ablation season hydrograph. Consequently, glaciers exert a strong influence over a catchment's runoff, even if glacier ice only occupies a small proportion of the catchment area.

Within a glacierized catchment, the drainage system that routes meltwater from the glacier surface to the ice margin can be conceptualized as a cascade. Meltwater can travel along flow paths at the surface (supraglacial), within the glacier's interior (englacial), and at the ice–substrate interface (subglacial). The structure, function and interaction of these systems determine the efficiency and transit time by which melt water travels from the point of melting to the glacier terminus.

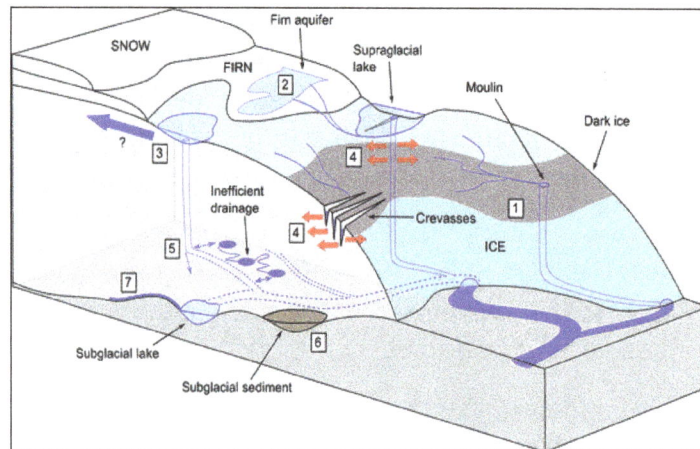

Conceptual model of glacier hydrology. Note the drainage takes an arborescent pattern in cross-profile, which is also apparent in plan view with a dendritic structure leading to the main subglacial conduit(s) or channel(s) at the glacier snout. Water flow is represented by blue arrows whose size infer discharge.

Accordingly, runoff from a glacier is controlled by the characteristics and hydraulic properties of these three drainage system components, which modulate both meltwater volumes and water quality through solute acquisition and sediment evacuation. Hydrological coupling between the supra-, en- and subglacial environments is fundamental in mediating the manner in which a glacier responds to environmental variability, primarily due to the role water plays in controlling ice motion. Here, focus is on mid-latitude, temperate valley glaciers, but for details on the hydrology of nontemperate glaciers and ice sheets.

Supraglacial Hydrology

The character of the supraglacial environment can be categorized into three key states, each with its own distinctive hydrology: snowpack, firn, and bare ice.

The Snowpack

The hydrology of snow in the supraglacial environment is generally the same as for other terrestrial snowpacks. Snowpacks on glaciers are the result of multiple snowfall events; in mid- and high-latitudes, these typically happen during winter months when little or no melt occurs, while for tropical glaciers, snow accumulates during warmer summer months. Consequently, glacier snowpacks are highly heterogeneous, exhibiting variations in crystal grain size and density from each snowfall event, leading to a complex, layered stratigraphy. After deposition, the snow structure is transformed by the pressure of overlying snow, by wind redistribution and compaction, and by internal processes of melting, refreezing, sublimation, and condensation. Such snowpack metamorphosis is driven by temperature gradients and snowpack age, and typically enlarges the crystal size or joins grains together (sintering), further influencing snowpack density. Combined, these processes result in local variations in snow density, ranging from 50 to 400 kgm^{-3}, and lead to contrasts in snowpack porosity and hydraulic conductivity, not only with depth but also over space.

As the melt season commences, snow surface meltwater begins to percolate into and warm the snowpack. The infiltration of liquid water and the latent heat released by refreezing begins to overcome the cold content of the snow, raising the snowpack temperature towards the melting point. Densification of the snowpack occurs as diurnal melt–freeze cycles cause the refreezing of percolating meltwater either within the pore spaces of the snow or as more massive ice lenses. Continued melt ripens the snowpack as gravity-driven percolation causes the wetting front to descend away from the surface into colder snow below. The percolation rate of meltwater can be approximated by Darcy's Law, with velocities of 10^{-4}–10^{-5} ms^{-1}. Critically, due to the structure and density variations within the snowpack, the wetting front is neither uniform nor horizontal, and preferential flow paths or flow fingers develop. The velocity of meltwater through a ripe snowpack is dependent on the physical structure of the snow and the water flux: greater volumes of melt travel more rapidly. As the snowpack becomes isothermal (at melting point) the increased water content raises its homogeneity and hydraulic permeability. Meltwater content and flow will be highest (and most concentrated) where melt rates are greatest. Drainage through a saturated snowpack is driven by the local slope and transport velocities can increase to 10^{-2} ms^{-1}.

Over the accumulation area of a glacier, at the peak of the summer melt cycle, the varied physical and thermal properties of the snowpack may be classified as three distinct snow facies: (i) dry snow is typically only found at very high elevations (or in the interior of ice sheets), where no melting occurs, the water content remains at 0%, and the snow makes no contribution to runoff; (ii) percolation snow occurs at intermediate elevations, where some summer melting occurs, meltwater (about 3% water content by volume) penetrates the snowpack or drains into the firn, but refreezing and storage occurs, ensuring melt does not contribute significantly to runoff; (iii) wet snow arises where summer seasonal melt raises the snowpack to 0 °C and vertical percolation results in saturated snow (8–15% water content) and lateral flow occurs, particularly at the glacier ice interface, or where ice lenses or layers occur. In contrast to cold or polythermal glaciers, the summer melt season warms the entire snowpack on temperate glaciers and the dry and percolation snow facies are not observed.

At lower elevations, in the ablation zone, the increasingly thin and saturated snowpack may become slush (>15% water content). Slush flows themselves may contribute to the removal of snow from the glacier surface. The slush zones commonly follow the seasonal retreat of the transient snowline and in this zone supraglacial lakes may form where glacier surface gradients are low and topographic depressions exist. Undulations in glacier surfaces can be formed by ice flow itself and by the topography of the glacier bed. Meltwater may also pond in crevasses during the early part of the melt season. Lake volumes of 10^4 m^3 have been described on valley glaciers, while on the Greenland ice sheet, supraglacial meltwater lake volumes of 10^7 m^3, with extents of up to about 9 km^2 have been reported. Supraglacial lakes are transient features usually forming early in melt season as supraglacial drainage is initiated, when saturated snow- or slush-plugs prevent flow through supraglacial stream courses. However, on ice sheets, lakes may persist between melt seasons.

At the end of the summer, the snowline position broadly corresponds to the equilibrium line altitude (ELA), which is the average elevation of the area on the glacier where accumulation equals ablation over a one-year period. At the ELA, slush and supraglacial lakes may persist during summer, but superimposed ice is also evident. Superimposed ice forms where meltwater refreezes at the glacier ice interface, most typically in early summer as meltwater percolates through the snowpack, but also in early autumn when residual melt draining from the accumulation area and precipitation refreeze as air temperature decreases. Although superimposed ice can form across the entire ablation zone, the high ablation rate at lower elevations removes evidence of its formation. Superimposed ice can reduce the permeability of surface glacier ice, so that accumulation of meltwater on the glacier surface is pronounced early in the ablation season.

The Firn

Firn represents the transition between snow and glacier ice, and is formed through the processes of firnification. This process represents the compaction and reorganization of snow and ice crystal structure, which in turn increases its density to >400 kgm^{-3}. Snow that survives over a full annual cycle becomes firn. The rate at which this snow transforms into firn (and ultimately into glacier ice) is dependent on the accumulation rate of snow, the temperature of the snow, and the presence of meltwater in the evolving snowpack. These controls impose a gradual reduction in the air or void space within the snowpack, and firn becomes glacier ice upon the occlusion of intercrystalline air passages to bubbles, at a density of about 830 kgm^{-3}. The density of firn increases with vertical distance from the snow surface and also with proximity to the ELA, where production and refreezing of meltwater is commonplace. These spatial trends towards an increasingly tight packed crystal structure result in a porosity gradient. Characteristically, firn is found above the ELA and its hydrology is important for runoff sourced in a glacier's accumulation area.

Due to the presence of firn, the accumulation area of a glacier represents a confined aquifer. Meltwater derived from the surface snow can percolate through the snowpack and into the firn layers. This meltwater drains through the higher porosity, unsaturated firn to form a saturated firn layer at depth, where the reduced permeability impedes further drainage. This saturated firn layer progressively develops following the onset of the melt season; the volume of water retained in this aquifer is dependent on, amongst other things, the snow melt rate.

The saturated firn layer on valley glaciers has been reported at depths of up to 15m below the firn surface, with thicknesses ranging from 1 to 7m above the firn–ice transition. The meltwater transit velocities of 10^{-4}–10^{-5} ms^{-1} in the saturated layer emphasize the firn's ability to delay snowmelt runoff at timescales of days to weeks. The slowed transport of melt water in accumulation area firn reservoirs results in the damping of any diurnal melt signal during summer. However, as the melt season ends, the saturated firn drains down-glacier or meltwater refreezes in situ (aiding firnification), at least partially reducing the firn aquifer's liquid water content.

The Bare Ice

During the melt season, the decrease in snowpack extent, and up-glacier recession of the transient snowline towards the ELA, results in exposure of glacier ice and initiate rapid supraglacial runoff from the ablation area. If unimpeded, this meltwater flows across bare surface ice at much higher velocities (typically 10^{-1} ms^{-1}) than those seen in the snowpack or firn.

On all glaciers during the melt season, the ablation area represents a transient thermal layer: in winter, the glacier ice surface is cooled below the pressure melting point (PMP), while during spring and summer the increasingly positive energy balance driving ablation ensures the seasonally cold surface layer initially approaches and then remains at the PMP. Because glacier ice is not a pure medium, the energy balance that drives melt is not uniform over space. Preferential radiation absorption at crystal boundaries or the presence of dust and impurities on the ice surface generates micro- (micrometer) to small- (centimeter) scale topographic variations. Meltwater may percolate through or along these flow paths and, by increasing flow volumes, create rills on the ice surface.

Incident solar radiation also penetrates the upper-most glacier ice, typically to a depth of about 2m. Through melting at subsurface crystal boundaries, the surface ice density can be reduced to about 550 kgm^{-3} creating a porous weathering crust through which water can be transported. Observations have suggested this shallow weathering crust zone behaves like a perched aquifer, with the capacity to store and release meltwater; however, this weathering crust aquifer is typically saturated, with its water table just a few centimeters below the ice surface. Local differences in ice structure (e.g. foliation) result in contrasts in hydraulic permeability in the ice, which may contribute to the formation of preferential flow paths within the weathering crust ice. Transport rates in the weathering crust are typically between 10^{-4} and 10^{-6} ms^{-1}.

By following the steepest surface gradient, surface rills coalesce to form supraglacial stream channels. Akin to terrestrial river catchments, supraglacial stream channels commonly develop dendritic or arborescent drainage patterns, with drainage densities ranging from <10 to >20 $km\ km^{-2}$, which generally decrease up-glacier where ablation is reduced. Due to the convexity of glacier ablation areas, elongated, subparallel drainage patterns are often observed. Similarly, the structural control imposed by down-glacier longitudinal foliation and annealed crevasse traces can result in rectilinear surface drainage systems. Delays of between one and eleven hours have been observed between peak melt production and peak supraglacial stream discharge; these delays relate to catchment size and geometry, but also the hydraulic properties of the weathering crust. The form of supraglacial flow paths and stream catchment areas varies glacier to glacier: drainage networks may be disrupted by crevasses occurring in response to site-specific glacier motion and the associated stresses in the ice.

The geometry of supraglacial streams is controlled by, amongst other things, stream catchment area and ablation rate. To accommodate increasing meltwater discharge, supraglacial stream flow velocity and water depth increases while channel width remains relatively constant.

Here, flow velocities of >$1ms^{-1}$ are common, even in small streams. Channel roughness is very low; the dimensionless Manning's "n" for supraglacial channels is typically <0.03. As is seen in terrestrial river systems, both meandering and step-pool morphologies can develop in supraglacial streams. Meander wavelength (with sinuosity of 1.0–2.5) relates to discharge and stream width, while the sinuosity of meanders is inversely associated with reach-length channel slope. Step-pool sequences appear to develop where channel slope exceeds about 15°.

The incision rate of supraglacial streams is principally related to channel slope and discharge. Viscous fluid friction within the turbulent flow can accentuate channel incision. Supraglacial stream water equilibrates to around +0.1 °C, meaning there is also sufficient thermal energy available to induce downcutting of ice-walled channels at rates of about $2mmh^{-1}$. On glaciers in temperate latitudes, typically the ice surface ablation broadly matches stream incision rates (about 0.1 $mday^{-1}$) and streams reform each melt season. Conversely, in colder climates, stream incision may be an order of magnitude greater than glacier ablation, resulting in the development of progressively more deeply incised "canyons" >2m deep and persistent over many years. The process of meandering will continue at depth and the planform surface expression of a supraglacial stream may not reflect its active channel.

As a result of the low permeability of surface and near-surface ice, the ablation area of a glacier exhibits a rapid hydrological response to changing melt rates. The rapid transport of meltwater via supraglacial streams in the ablation area ensures that the diurnal melt signal is apparent in the meltwater discharge delivered from the bare-ice of glaciers. However, the configuration of drainage structure in the supraglacial environment is, in part, dependent on the occurrence of features that provide hydrological access to the englacial environment.

Englacial Hydrology

Massive glacier ice is permeable, allowing englacial water to permeate through it slowly. With a density of 830–910 kgm^{-3}, at the microscale, glacier ice crystals are surrounded by a film of liquid water; these films form a network of veins at the junction between three or more individual ice crystals. Typically 0.1–10 μm in diameter, the size of the veins depends on the ice crystal size, meltwater solute content, pressure, and temperature. Ice crystals range from <1mm to 10 cm in diameter and the size of the vein junction correlates positively with crystal size. The process of ice crystal formation itself rejects solutes and impurities, resulting in solute-rich (or saline) water within the intercrystalline or interstitial vein network. The lower freezing point of saline water compared to that of pure water allows for the existence of larger veins. Increases in overburden pressure reduce the PMP of ice, potentially enlarging the vein network in ice nearer the glacier bed; the hydraulic permeability of the ice in the lowermost several tens of meters in a glacier increases. Englacial strain heating resulting from the flow of the ice can also contribute to interstitial meltwater (0.01% by volume). However, irrespective of crystal size and solute content, as ice temperatures decrease, the vein network contracts; the porosity and permeability of ice at the PMP is considerably higher than for cold ice. Liquid water can be present in ice substantially below the PMP, but usually <0.1% by volume, while the water content of glacier ice at the PMP may rise to several percent by volume. The flow in the vein network is small, however, with laboratory studies suggesting velocities of between 10^{-8} and 10^{-13} ms^{-1}. The presence of air bubbles and debris, capillary forces, the deformation and recrystallization of ice, as well as high confining pressures, are all likely to impede this flow.

Despite the low rate of water transport, the interstitial vein network is prevalent throughout a glacier and water will generally flow towards the glacier bed. Assuming ice is a porous medium, water moves from high to low hydraulic potential (Φ). Hydraulic potential is the sum of the gravity-defined potential energy and the water pressure defined by the force imposed by the overlying ice. Theoretically, water flows perpendicular to the 3-D plane with equal hydraulic potential – the equipotential surface – and follows the steepest potential gradient. On valley glaciers, equipotentials typically dip up-glacier at 11-times the ice surface slope. Using calculations based on glacier geometry alone (e.g. ice extent, topography, and thickness) it is possible to calculate an informative overview of likely flow paths in the englacial environment.

Water flow, directed by the hydraulic potential, generates heat proportional to the water flux, further enlarging already large veins in the ice. As such a vein enlarges, its water pressure drops compared to smaller veins. Water flow will be directed towards the larger evolving conduit and, consequently, large conduits develop and grow at the expense of smaller flow paths. This results in the development of an arborescent drainage pattern following the gradient of equipotentials through the englacial environment. However, studies of flow paths in the englacial environment from direct observations (glacier speleology), ice-penetrating radar, and video imagery from boreholes do not appear to fully support the theoretical englacial drainage system as driven by hydraulic potential. Englacial conduits following subhorizontal pathways, vertical descent depths of up to 60m interrupted by small steps and horizontal galleries, or small (<0.1 m) englacial drainage structures with shallow dipping orientations have contributed to a growing realization that structural controls (foliations, fractures, variations in ice rheology, etc.) may provide a greater influence on the geometry of drainage features, and local hydraulic potential fields, than was previously appreciated.

Crucially, in view of the limited water transfer capable by the primary permeability of englacial ice itself, it is the larger-scale structures (e.g. crevasses, fractures, and moulins) that provide the pathways for the bulk of seasonal meltwater to access the englacial environment, even if hydraulic potential governs the general direction of water flow. These macroscale flow paths that represent secondary permeability may allow transit velocities of between 10^{-1} and $1 \mathrm{ms}^{-1}$. Most commonly, crevasses exist as direct pathways to the glacier interior. Such crevasses form in response to high tensile strain rates and have depths of up to about 30m in temperate ice (ice at the PMP), and greater in colder, more brittle ice. Crevasses truncate any supraglacial streams they intersect and sequester water from them. Energy dissipation by flowing water at the crevasse base may develop a vertically descending conduit at depth. The stress field within the ice causes a lateral deviation of such evolving conduits, with observations suggesting a down-glacier orientation of about 25° from vertical. Where ice flow causes crevasses to close, but supraglacial drainage continues to incise a notch into the up-glacier edge of the former crevasse, a vertical shaft (moulin) forms. Moulins from <1m to 10m in diameter may be long-lived features, persisting for multiple melt seasons.

Where crevasses or moulins become waterfilled, the process of hydrofracture may arise. Here, the mass of meltwater promotes fracture extension at the base of the crevasse or moulin. The depth of the hydrofracture depends on the volume of stored water and the persistence of meltwater delivery to the fracture locus.

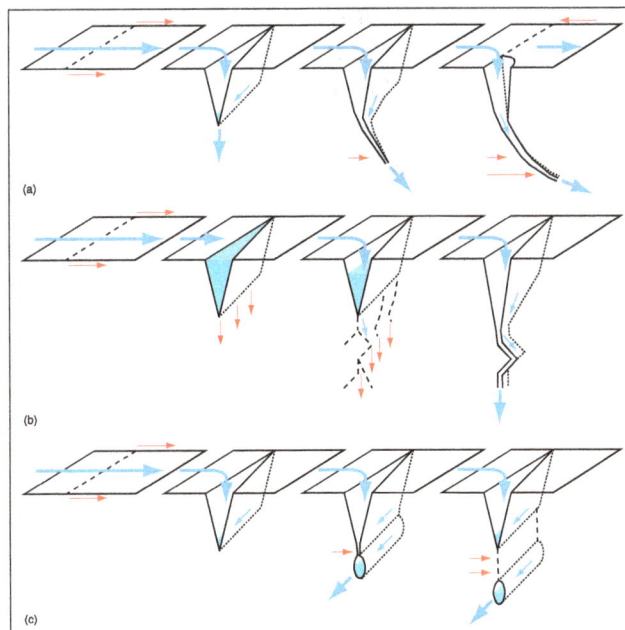

Illustration of hypothetical evolution of englacial conduits from crevasses and meltwater streams. (a) Classical moulin formation from truncation of a supraglacial stream; (b) crevasse hydrofracture promotes englacial drainage path; (c) crevasse-bottom incision and glacier motion isolate an active englacial channel; (d) cut-and-closure formation of englacial channel, where meandering at depth ensures englacial conduit geometry does not reflect original supraglacial stream path. Water flow is represented by blue arrows, and forces acting on ice shown are by red arrows. Ice flow is left to right for (a) to (c) and from back to front in (d).

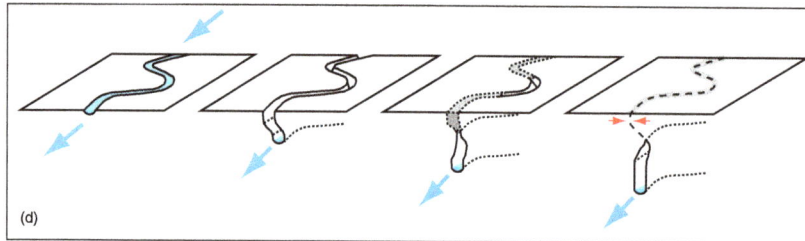

In the absence of a continuous water flux to the fracture base, refreezing will reduce the likelihood of the fracture descending to greater depths. Under optimal circumstances, crevasse hydrofracture has been observed in glaciers and ice sheets to propagate to depths of 10^2-10^3 m in short (hours, days) timescales. As with crevasse and Moulin drainage path formation, the changes in effective stress field with increasing depth can result in the hydrofracture flow path becoming inclined at progressively shallower angles. Hydrofracture is also not limited to a vertical orientation and appropriate water pressures and structural weaknesses can facilitate shallowly dipping drainage path formation.

It is possible for supraglacial streams themselves to develop into englacial conduits, especially where stream incision exceeds general ice ablation rates: snow bridges and metamorphosed and deforming ice can cover these incised flow paths, resulting in enclosed, high-roofed englacial channels called cut-and-closure channels. The geometry of these channels resemble supraglacial streams, with flow generally being at atmospheric pressure. Processes of vertical incision and knick-point retreat rates of 0.1–0.5 mday^{-1} in these channels can lead to the formation of moulin-like forms at the channel head. Similarly, subhorizontal channel forms have been suggested to arise from the entombment of conduits formed at the base of the saturated snow zone in the accumulation area of ice masses. Crevasses may also lead to subhorizontal drainage structures: water flow in crevasse bottoms may initiate a drainage route, and subsequent crevasse closure can isolate the drainage structure from the supraglacial environment. Speleological investigations have also revealed that ice structures – such as refrozen or annealed crevasse and fracture traces, debris bands, thrust faults, and highly permeable ice – all provide locations of high hydraulic transmissivity, supporting conduit inception. Studies have reported englacial fractures can be found throughout 96% of a glacier's depth. Fractures in ice can also result from frictional drag at the glacier bed, with basal crevasses propagating vertically upwards into the englacial environment.

Current wisdom suggests englacial drainage flow paths close up over winter months and reopen each summer, when water pressures decrease with channel enlargement. The closure rate of water-free ice-walled channels is a function of ice overburden pressure and viscosity, largely controlled by temperature. In cold ice, even small channels can persist for over a year. In contrast, temperate ice deforms rapidly and, once empty of water, conduits close rapidly: theoretically, at a depth of 100 m, a conduit's diameter may halve in <20 days. However, englacial conduit closure rates are substantially reduced if they are water-filled, as the stored water impedes ice deformation. The englacial system, therefore, can represent a reservoir with capacity to store considerable volumes of water, particularly over winter months. Empirical water balance studies have demonstrated disparities between input and output meltwater volumes that suggest such processes of storage and release occur within the englacial hydrological system.

Subglacial Hydrology

Assuming the glacier bed is at the PMP, water reaching the ice-bed interface flows towards the glacier terminus, orthogonal to the equipotential contours formed by the intersection of the equipotential surfaces with the bed. Water can exist at the ice-bed interface beneath cold ice areas due to the hypersalinity of mineralized waters resulting from solute rejection during melting and refreezing. Meltwater at the glacier bed may also be sourced from basal and frictional heating in areas of temperate ice, although annually this rarely exceeds about 10^{-1} m of melt. Once present at the ice-bed interface though, meltwater flow is complicated by variability in both supply and bed conditions.

Where ice is thin, channel enlargement by viscous heat dissipation may offset channel closure, and such channels would be open with meltwater flowing at atmospheric pressure. In this situation, bed geometry is the primary factor in defining the orientation of subglacial channels. However, observed channel locations have often appeared to mirror those reconstructed from hydraulic potential assuming water at ice-overburden pressures. To reconcile this dichotomy, it is quite possible that drainage pathways close by deformation in winter months and subsequently reopen each summer. Since conduits are water-filled, and therefore at (or temporarily above) ice-overburden pressure during re-formation in the spring, their initial location is controlled by hydraulic equipotential pressures. Subsequently, and despite reduced water pressures, melt rate is insufficient to allow lateral migration to the location dictated by theoretical open-channel flow (i.e. the low point of the glacier bed). Nonetheless, there is a broad agreement that water flow at the bed occurs in one or both of two qualitatively and hydraulically different conceptual flow systems: distributed or "slow" flow and discrete or "fast" flow.

Distributed (Slow) Subglacial Drainage

Distributed subglacial drainage occurs through spatially extensive, non-arborescent (or anastomosing) flow pathways at relatively slow velocities of $<10^{-2}$–10^{-3} ms^{-1}. For a glacier with a hard bedrock sole at the PMP, microscale pressure-driven melting and refreezing (regelation) occurs around bedrock protuberances up to about one meter in length. The meltwater generated by this process flows around bedrock bumps and clasts as a millimeter-thick film. Sedimentological analyses reasoned that the lack of particles <0.2mm at the rock–ice interface demonstrated the presence of water films capable of eroding finer particles, while numerical analyses indicated an upper limit to film stability of about 4 mm. In reality, small-scale bed roughness probably reduces stable film thickness to substantially less than this. Raised viscous heat dissipation in thicker films, or local increases in water discharge due to supply or bedrock variability, result in the development of preferential drainage paths and protochannels. Where ice flow is sufficiently fast, ice may separate from the bed in the lee of a bedrock protuberance, leading to the development of cavities. Subglacial meltwater may collect in, and thermally enlarge, cavities. The initial cavity size is dependent on bed topography, ice velocity, and basal shear stresses. Cavities may be either autonomous or interconnected by narrow channels (orifices) where water flow is sufficient. A network of interlinked cavities forms an anabranching linked cavity system. For small discharges, ice melt is unimportant for maintenance of the cavity system, so small increases in water pressure lead to a greater carrying capacity of the system. However, if meltwater discharge through the inefficient linked cavity system increases further, preferential channel-like flow paths can develop.

Many glaciers are not entirely in direct contact with consolidated bedrock. Instead, subglacial investigations and observations of recently deglaciated surfaces commonly indicate the presence of unconsolidated sedimentary deposits. With a hydraulic conductivity typically in the range 10^{-6}–10^{-12} ms^{-1}, Darcian porewater flow is likely through subglacial sediments, although extremely difficult to measure directly. However, macroporous or preferential flow paths may develop in subglacial till, and porewater flow is likely to occur in conjunction with other more efficient drainage systems.

Discrete (Fast) Subglacial Drainage

Discrete subglacial drainage occurs through channels or conduits, typically forming a stable arborescent network along which meltwater flows rapidly at velocities of 10^{-1}–10^{-2} ms^{-1}. The relative instability of distributed drainage systems to increased water flux means that they break down as water discharge accumulates either gradually down-flow or locally due to point inputs. Here, preferential drainage paths develop by melting the overlying ice to form semi-circular subglacial channels, Röthlisbergeror R-channels, which follow the hydraulic gradient. However, water pressures measured in subglacial channels have been found to be higher than those calculated for ideal R-channels, indicating unusually rough or sinuous conduits, or a cross-sectional shape that is broad and low rather than semi-circular. In these Hooke- or H-channels, melting is concentrated on the conduit walls, since channels are not continually water-filled and lateral closure is limited by friction with the bedrock. Where bedrock may be readily eroded or where major subglacial channels persist for sufficient time, such channels may incise into the rock rather than the ice, forming Nye- or N-channels.

In the case of a poorly consolidated sediment bed, due to the instability of films and porewater flow to externally driven variations in water flux, it is thought that shallow anabranching canals incise into the deformable bed material. Canals are enlarged by water flow removing sediments, with sediment creep and failure of the canal wall counteracting the tendency for growth.

Given the spatially restricted nature of discrete subglacial drainage and the spatially extensive nature of distributed subglacial drainage, it is anticipated that the two systems likely coexist side-by-side. Discrete channels are likely to form at point sources, such as beneath Moulin delivery points, and to extend down-flow of them, irrespective of their location. Channels will also generally increase in representation down-glacier, where accumulated subglacial meltwater flux is largest. In between these channels, distributed drainage will dominate. Indeed, at temperate glacier beds at the PMP, the distributed component will be ubiquitous between discrete channel pathways. Critically, evidence suggests not only that such a combination of subglacial drainage systems do coexist, but that they interact at a variety of scales. For example, coupling between porewater flow and channelized flow has been observed whereby high daytime water pressures within channels drives water from them into the surrounding distributed system. At night, when the channel water pressures are markedly lower, that water returns, transferring solute and suspended sediment from the surrounding locality to the channel.

Temporal Evolution of Glacier Hydrology and Runoff

Conceptually, spatial and temporal changes in temperate glacier hydrology are thought to involve, on an annual cycle, a glacial drainage system that closes over in winter and reopens during summer months.

The relatively rapid rate of ice creep in temperate glaciers, with ice at the PMP, makes it unlikely that englacial and subglacial conduits or fast drainage structures remain open once devoid of liquid phase meltwater during winter. However, once closed, meltwater can become trapped through the winter in cavities at the glacier bed. At the commencement of the melt season, the volume of water stored within the glacier initially increases due to poorly interconnected drainage structures (e.g. snowpack, isolated crevasses, and flooded moulins) and the presence of a slow drainage system at the glacier bed. As surface warming continues through the spring and early summer, the surface snowpack becomes isothermal, and snowmelt floods (spring events) occur, which flush a large volume of meltwater through the glacier drainage system, resulting in an early melt season peak in runoff volumes (ans commensurate ice velocity). This spring event may also release previously stored meltwater. As the summer progresses, ablation is heightened across the glacier surface and increasing volumes of water are routed into the glacier interior. This increases water pressures and destabilizes slow drainage structures such that a channelized, fast flow network develops at the expense of distributed drainage. The hydraulically efficient system then probably persists for the remainder of the melt season with sufficient water flux to maintain fast drainage structures. Once ablation and meltwater volumes decline, the discrete, fast flow drainage structures begin to close through ice deformation.

Importantly, these temporal changes do not occur uniformly across the entire glacier bed but have a systematic spatial expression that is driven by the up-glacier expansion of the area of intense surface melting. This area corresponds to the expanding area of bare ice exposed by the melting of the (hydraulically buffered) supraglacial snowpack. The retreat of the snowline towards the ELA progressively exposes moulins and crevasses that become hydraulically active, delivering melt to englacial or subglacial drainage flow paths. Consequently, water pressures within and at the bed of a glacier are not spatially uniform, and the growth of a fast, discrete drainage system evolving from a slow system follows the retreat of the snowline up-glacier. The growth in extent of an efficient drainage system is also likely to reduce storage within the glacier, with a greater degree of coupling between melt processes and runoff volumes following the reduction in both the snow/firn aquifer and impedance of subglacial flow. Meltwater discharge as the snowline retreats to higher elevations has been demonstrated to result in an increase in bulk meltwater runoff volumes as well as a reduction in the time lag between peak temperatures and peak runoff. As the season progresses, the diurnal runoff signal becomes increasingly peaked due to the loss of hydrological characteristics capable of damping the transport of melt waters. Later in the melt season, with falling rates of ablation, the reduced supply of melt water is insufficient to maintain the fast drainage structures, which begin to close and decrease in extent as they transition back to, and integrate with, the distributed flow structures.

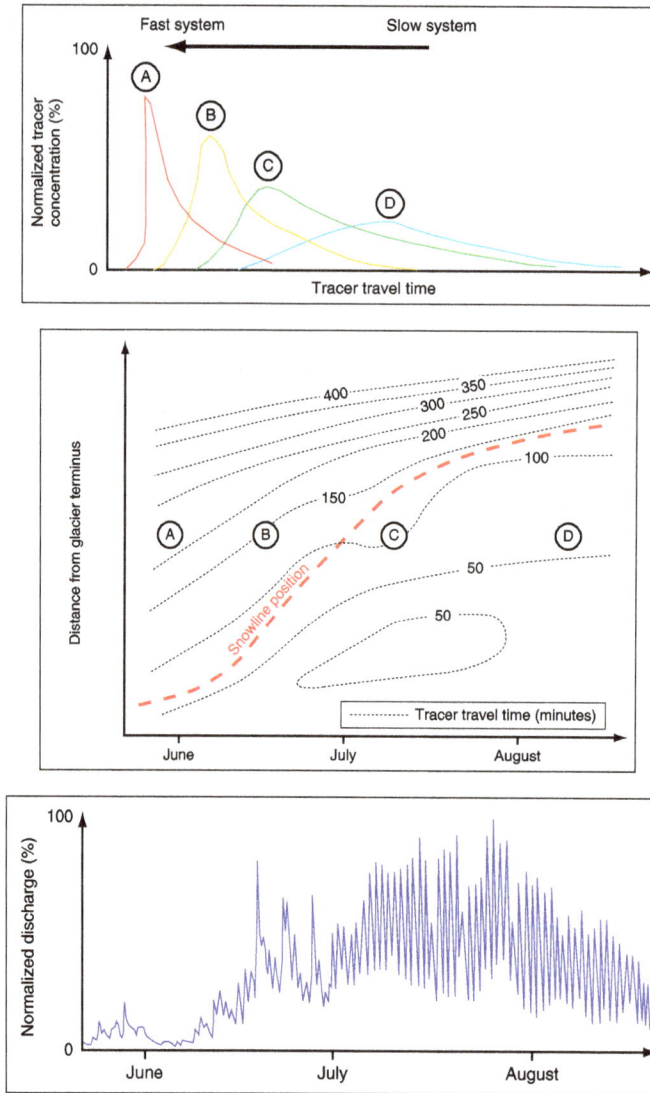

Schematic of the spatial and temporal variation in meltwater travel to the proglacial environment. (a) Systematic change in tracer transit over the melt season, showing decreasing dispersion and increasing velocity; (b) the seasonal, up-glacier extension of fast meltwater transit times mirroring the retreat of the snowline; (c) the associated seasonal runoff hydrograph. The temporal positions of idealized tracer breakthrough curves (A–D) are shown.

The temporal and spatial progression of a temperate glacier's hydrological system imparts systematic changes in the runoff hydrograph. Runoff may flow from a glacier throughout the year, with a base flow associated with slowly draining water, volumes of which only change at longer (most typically seasonal to annual) timescales. From the onset of melt, and following any snowmelt flood events, a glacier's runoff hydrograph typically exhibits rising discharge volumes and increases in the amplitude of diurnal discharge variations as the drainage system progressively becomes more evolved and efficient. The time lag between peak melt and peak discharge at the glacier terminus is progressively reduced, and may be less than one hour. If melt season snowfall occurs or synoptic conditions reduce melt rates, discharge (and diurnal variations thereof) are commonly diminished. As the melt season ends, the hydrograph typically shows a rapid decline in runoff volumes.

Quality of Glacial Runoff

For most glaciers, water emerges from the glacier terminus in a few discrete proglacial streams associated with the dominant fast subglacial flow paths. The water quality in these streams is dependent on the flow paths meltwater has followed; a glacier's hydrological structure is critical to the acquisition of solutes (ions dissolved in water) and entrainment of suspended sediment by waters emerging at the ice margin.

Solutes

The hydrochemistry of meltwaters within glacierized catchments has been studied since the 1970s. The primary hydrochemical constituents in glacial meltwaters are solutes derived from atmospheric deposition and the weathering of catchment bedrock and glacial sediments, predominantly in subglacial or ice-marginal environments. The complex system of solute acquisition by glacial meltwaters may be conveniently simplified by a two-component model whereby bulk meltwater is composed of "quick" and "delayed" flow. Conceptually, the former can be considered as rapid supraglacial and englacial flow, along with channelized subglacial flow, which is largely devoid of solutes, while the latter relates to slower subglacial transport pathways, enriched in solute. In practice, however, these components do not exist as discrete entities, and solutes are acquired to varying degrees by glacial meltwaters that follow composite pathways, seriously undermining the mixing model approach. However, it is well ascertained that solute acquisition is influenced by the residence time of waters at the ice-bed interface. In response to the variation in meltwater production during the summer melt seasons, typically high solute concentrations are observed at times of low discharge, with low flux and long contact times, and low concentrations during times of elevated meltwater discharge, with high flux and shorter contact times, both at the diurnal and seasonal timescales.

Precipitation in glacierized catchments is the primary source of base levels of solutes observed in runoff. Seasonal snowfall and snowpacks are, therefore, important as sources and stores of solutes within glacierized catchments. As snow begins to melt, the percolation of meltwater results in changes in that snowpack's chemical composition. Field and laboratory studies have shown that 80% of the snowpack solute load may be released within about the initial 25% of meltwater runoff. This leaching or elution of solutes may be complicated by snowpack heterogeneity. Seasonal elution has also been observed in runoff from firn. This removal of solutes from snowpack and firn, the sources of glacial ice, means the ice itself is relatively dilute. As the snowline retreats up-glacier during the ablation season, dilute ice-melt comes increasingly to dominate supraglacial runoff.

The relatively high solute concentrations measured in proglacial meltwaters must, therefore, be acquired by chemical weathering of sediments in ice-marginal or subglacial environments. It is now widely recognized that hydrolysis and carbonation reactions dominate rock–water interaction and chemical weathering within glacierized basins. Calcium ions (Ca^{2+}) are the dominant cation in glacial runoff, which reflects its relatively rapid rate of dissolution from silicate and carbonate rocks, although silica concentrations tend to be lower than in nonglacial runoff because silicate weathering is depressed in cold subglacial and ice-marginal environments. The acquisition of other base cations depends on a catchment's specific geology and geochemical susceptibility. However, observations of persistently elevated sulfide ion (SO_4^{2-}) concentrations suggest alternative weathering reactions occur along subglacial flow paths. Anaerobic nitrate-reducing or sulfate-reducing microbes and oxidizing chemotrophs have been shown (or inferred) to exist in a wide range of glacial environments, and subglacial chemical weathering can be increased up to eightfold through microbially catalyzed reactions.

Sediment

The runoff from glacierized catchments characteristically exhibits high suspended sediment loads that influence the use and management of that water. Consequently, substantial research attention has been devoted to glaciofluvial sediment transport, which is typically confined to a limited melt season. Analyses have shown that annual specific glaciofluvial sediment yields, which routinely exceed 102 t km^{-2} $year^{-1}$, are substantially greater than global averages for other terrestrial catchments.

In view of the conceptual model of valley glacier hydrology during the ablation season, the critical distinction between the quick flow paths of the largely debris-free supraglacial and englacial systems, with low or intermediate suspended sediment concentration (SSC), and the slower, delayed flow through the subglacial drainage system, exhibiting high SSC through entrainment of rock flour and subglacial debris, is important. The transfer of fine-grained rock flour in suspension dominates sediment evacuation from most glaciers, and particularly from temperate-based glaciers. The concentrations of sediment transported via supraglacial drainage paths, although typically <0.5 g l^{-1}, are highly dependent on the nature and extent of debris at the glacier surface and within the ice body as well as delivery of sediment derived from extraglacial locations, such as lateral moraines, to the supraglacial environment. Debris may be released from englacial and basal ice through the melting and enlargement of conduits by viscous heat dissipation. However, where the glacier bed is composed of unconsolidated materials, the entrainment and removal of sedimentary products will be the dominant source of suspended sediment entrained in meltwaters passing through subglacial drainage paths, with SSC commonly ranging between 1 and 102 g l^{-1}. Entrained sediment may itself increase SSC through mechanical erosion (abrasion) of bedrock or basal sediments. However, the relationship between SSC and discharge is neither linear nor stable in proglacial rivers and sediment yields can vary markedly at diurnal, seasonal, and annual timescales.

Changes in the rate of meltwater production at the glacier surface and the hydrological configuration within a glacier control the temporal variability in suspended sediment transport by influencing the processes of erosion, entrainment, and transfer. The volume of sediment evacuated by runoff is typically viewed as a function of the availability of debris at the subglacial ice-bed interface. At a seasonal timescale, the sediment yield relates to the evolution of the subglacial drainage system: SSC (sediment evacuation) can be high during the spring melt, when snow-melt flushes the glacier's subglacial drainage network and destabilizes the slow drainage system. Subsequently, up-glacier retreat of the snowline and corresponding growth of the fast subglacial drainage system may cause short-term increases in sediment delivery as the basal area accessed by subglacial meltwater increases. As the melt season progresses and diurnal melt cycles become more pronounced, sediment concentrations may be elevated as subglacial sediments are eroded by enlarging fast drainage structures. However, decreases in suspended sediment concentration may arise on a seasonal basis, arising from the exhaustion of available fines, as the subglacial drainage system spatially stabilizes and then begins to close during the latter portion of the melt season.

Short-term variability in sediment transport by runoff is usually dominated by a marked diurnal cycle, typically characterized by a strongly positive relationship between discharge and SSC. However, the direct, positive association between SSC and meltwater discharge commonly breaks down. On the diurnal timescale, clockwise hysteresis is frequently observed, whereby the flushing and exhaustion of sediment on the rising limb of the diurnal hydrograph occurs, and there is significantly less sediment available for transport when discharge declines. Similarly, anticlockwise hysteresis may occur in cases where subglacial channel water pressures reduce on the falling hydrograph limb and return flow from the distributed system elevates the SSC entrained at lowered discharges. Shorter-term subdiurnal variations in SSC, so-called sediment pulses, have been related to sudden changes in drainage system connectivity over the glacier bed, to glacier dynamics that result in reconfiguration of the drainage system, and to snowmelt or rainfall events where the drainage system becomes inundated. Additional influences on glaciofluvial sediment yields from glacierized catchments may involve the reworking and erosion of typically unconsolidated materials associated with moraines, and braid plains, or sandurs that typify ice-marginal environments and contemporary proglacial areas.

Chapter 4

Ecohydrology and Isotope Hydrology

The branch of science which seeks to study the interactions between the different ecological systems and water is termed as ecohydrology. Isotope hydrology refers to the branch of hydrology which is involved in estimating the origins and age of water through isotope dating. The topics elaborated in this chapter will help in gaining a better perspective about ecohydrology and isotope hydrology.

Ecohydrology

Ecohydrology is a subdiscipline of hydrology focused on ecological aspects of hydrological cycle.

The increasing human population has resulted in increasing pollution and degradation of water and biogeochemical cycles at a basin scale. Declines of water quality and biodiversity at a global scale are sobering evidence that the prevailing hydrotechnical approach in water management, focused on sewage treatment plants and regulation of hydrological processes such as floods and droughts, are crucial but not sufficient. To achieve sustainable water management it is necessary to develop a complementary new approach. This new approach should be based on the recent progress made in understanding the functioning of aquatic and terrestrial biota.

Ecohydrology Key Hypotheses

- Hypothesis H1: "The regulation of hydrological parameters in an ecosystem or catchment can be applied to control biological processes".

- Hypothesis H2: "The shaping of the biological structure of an ecosystem(s) in a catchment can be applied to regulate hydrological processes".

- Hypothesis H3: "Both types of regulation integrated at a catchment scale and in a synergistic way can be applied to the sustainable development of freshwater resources, measured as the improvement of water quality and quantity (providing ecosystem services)".

It should be stressed that according to the ecohydrology concept, the overall goal defined in the above hypotheses is enhancement of ecosystem carrying capacity for ecosystem services and resilience, resilience to anthropogenic stress.

Ecohydrology Principles

- Hydrological (framework): Integrating of understanding water and biota interplay at a catchments scale (GIS) for identification of threats and opportunities for sustainable water, economy and society. It covers such aspects as:

- ○ Scale: Mesocycle water circulation in a basin - the template for quantification of ecological processes;

- ○ Dynamics: Water and temperature - the driving forces for terrestrial and freshwater ecosystems;

- ○ Hierarchy of factors: Abiotic factors to hydrology dominate, however, when these are stable and predictable, biotic interactions begin to manifest themselves.

- • Ecological (target): understanding of the evolutionarily established resilience and resistance of ecosystems to stress and patterns of ecological succession to increase their carrying capacity against human impacts.

- • Ecological engeneering (methodology): using ecosystem properties as a management tool for restoration of biodiversity, improve water quality and enhancement of ecosystem services for society by:

- ○ Dual regulation of hydrology by shaping biota and, vice versa, regulation of biota by altering hydrology.

- ○ Integration: at the basin scale various types of regulations should be integrated towards achieving synergy to stabilize and improve the quality of freshwater resources at a basin scale.

- ○ Harmonization of Ecohydrological measures with necessary hydrotechnical solutions (e.g. sewage treatment plants, levees in urbanized areas, etc.).

Ecohydrology of Terrestrial Ecosystems

Water controls the dynamics of terrestrial ecosystems directly, as a resource for the biota, and indirectly, as a driver for abiotic processes on the Earth's surface, in the atmosphere, and belowground. The biota in turn, modulates several hydrological processes and the rate of the water cycle. Most terrestrial vegetation interacts with hydrological processes through the soil-water balance, which is affected by soil properties, random climate drivers, and feedbacks with the biota. River flow enhances the ecohydrological connectivity of the landscape, spreading sediments, nutrients, propagules, and waterborne disease through waterways.

Terrestrial Vegetation Plays a Crucial Role in the Water Cycle

It modulates and sustains evapotranspiration and precipitation. Without vegetation the terrestrial water cycle would be much slower because of smaller evaporative losses and lower precipitation rates. Land regions lose their rainwater input either as evapotranspiration or as surface and groundwater runoff. Evapotranspiration has vividly been termed "green water flow", whereas runoff and groundwater fluxes have been referred to as "blue water flow". Root-zone soil moisture is the major stock of green water, whereas blue water contributes to stream flow and aquifers. Both green and blue water have important ecological functions. In particular, transpiration is coupled with photosynthesis and plant primary production through stomatal regulation, which modulates water vapor and carbon dioxide (CO_2) exchanges with the atmosphere. With some notable exceptions (e.g. phreatophytes and hydrophytes), green water supports most terrestrial vegetation and food production because most natural ecosystems and rain-fed agroecosystems rely on soil moisture uptake. Blue water flows provide water and other resources to a number of organisms and societies. They also enhance landscape connectivity, which is fundamental to the transport of nutrients and sediments; the spread of species, populations, and pathogens; and the maintenance of biodiversity. Green and blue water reservoirs are the staging ground of many biogeochemical processes that are fundamental for life on Earth.

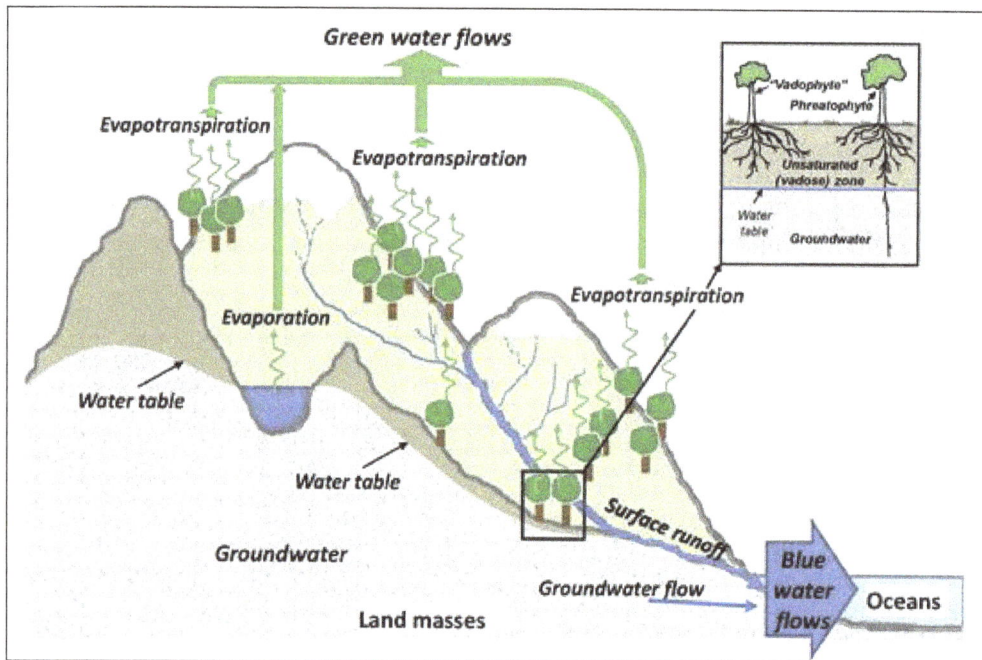

Schematic representation of the major water fluxes from terrestrial landscapes. Water leaves land masses as surface or groundwater flow into the oceans (blue water flow, approximately 39% of the precipitation input on land) and as water vapor flow (i.e. evaporation and transpiration) into the atmosphere (green water flow, approximately 61% of precipitation). Irrigation increases evaporative losses (green water flows) at the expense of blue water flows. Lakes, rivers, and groundwater are blue water stocks. Most terrestrial vegetation (except hydrophytes and phreatophytes; see inset) relies on water from the unsaturated zone (green water stocks). Blue water flows contribute to landscape connectivity and are vectors of sediments, pathogens, and propagules.

The study of the interaction of the water cycle with biota has been recently termed ecohydrology. This designation was initially used to denote an integrated study of ecological and hydrological processes in wetlands. The same approach was later extended to terrestrial ecosystems and to the study of the relationships between freshwater flows and ecosystem services. Ecohydrology is an engaging new discipline that is drawing the attention of an increasingly broad segment of the science community and spurring the development of new academic programs. What is particularly exciting and new about this field? Ecohydrology is contributing to the development of unifying concepts and new, testable theories for the understanding of complex patterns and processes in ecosystems. Here we review some of these concepts in the context of green water and blue water interactions with the biota.

Ecohydrology of Green Water Flows

Ecologic Function of Soil Water

Soil moisture affects biota directly, by controlling the availability of resources for organisms, and indirectly, by modifying abiotic processes that affect ecosystem dynamics. Plants and soil micro-organisms respond to conditions of limited soil moisture availability by slowing their rates of transpiration, photosynthesis, microbial decomposition, and mineralization. These processes are also limited when soils are nearly saturated, whereas other microbial processes, such as denitrification and biogenic emissions of methane, are enhanced in soil microsites undergoing anaerobic conditions. In dryland regions, mechanisms of competition among plant and microbial species also depend on soil moisture. Soil moisture may also indirectly affect the biota through its impact on abiotic processes such as surface runoff and water erosion, which contribute to the spatial redistribution of soil resources. In complex terrain, soil moisture patterns may exhibit geomorphically controlled heterogeneities, which induce spatial variability in species distribution and in the rates of processes such as photosynthesis and soil respiration.

Biotic Controls on Soil Moisture Dynamics

Green water stocks and flows are affected by biological processes. On the one hand, canopy and litter interception, root uptake, and transpiration are directly influenced by vegetation, and soil infiltration capacity and runoff are affected by soil bioturbation associated with plant roots, biological crusts, animal burrowing, and trampling by wildlife and livestock grazers. On the other hand, terrestrial ecosystems modify land-surface attributes such as albedo, roughness, and depth of active soil, thereby affecting surface energy and water-vapor fluxes, boundary layer dynamics, precipitation, and soil moisture regime.

Effects of Environmental Change

Changes in atmospheric composition, climate, land use, and rates of biological invasion have a profound effect on the interaction of hydrologic processes with the biota. The response of terrestrial ecosystems to global environmental change is often mediated or sustained by changes in soil moisture dynamics. Increased levels of atmospheric CO_2 enhance ecosystems' water-use efficiency, with possible important impacts on the soil moisture regime. Woody plants sometimes encroach upon arid grasslands, leading to a decrease (or sometimes an increase) in runoff and stream flow and to an increase in bare-soil evaporation. Climate warming enhances evapotranspiration, thereby accelerating the hydrologic cycle, though this effect may be partially offset by atmospheric aerosols. Anthropogenic disturbances affect nearly all ecosystems on Earth, and the resulting changes in land use and land cover modify the dynamics of key hydrological processes. For example, the replacement of forest vegetation with croplands and pasture typically reduces soil infiltration capacity, root depth, and surface roughness, and increases albedo. These changes in land-surface properties lead to decreases in evapotranspiration, precipitation, and water table depth, and increase runoff. The opposite effects occur in the case of afforestation. One of the major impacts of land use on the water cycle is associated with irrigated agriculture, which further disrupts the natural balance between green and blue water flows. Irrigation-induced reductions in river flow and man-made structures such as dams have major implications for ecohydrologic connectivity. Moreover, the greater soil evaporation in irrigated areas often enhances salt accumulation and dryland degradation. In many wetland environments land-use change and climate warming cause the drying of historically waterlogged soils, thereby increasing the rates of soil respiration and turning these landscapes into carbon sources.

Plants as Mediators of Green Water Flows

Ecohydrology aims to understand how vegetation contributes to green water flows. This research has focused on mechanisms controlling sap flow, water uptake through roots and shoots, and stomatal regulation, as well as on these mechanisms' interactions with hydrological processes driving the soil water balance.

The commonly accepted mechanism of water ascent in the plant xylem, derived from the cohesion-tension theory, recently has been confirmed by detailed laboratory experiments, despite the challenge that mechanisms other than root-to-shoot water potential gradients might contribute to water ascent. Such a controversy highlights that advances in monitoring microfluidic processes in plants are necessary to improve our understanding of plants' water and nutrient uptake.

Similarly, in the past two decades the study of root structure and function has advanced the understanding of interspecific facilitation and competition for soil moisture. More recently, it has been shown that mechanisms of "hydraulic lift" and downward redistribution have an important impact on the way deep-rooted plants contribute to green water flows. The inclusion of these mechanisms in some models of soil moisture dynamics and atmospheric general circulation has shown how hydraulic redistribution enhances dry-season transpiration and modifies the seasonal climate. In some species, water uptake through direct absorption of moisture by the shoot has also been reported as an alternative to the common paradigm of root uptake.

Ecosystem-level studies have provided new insights into the dynamics of plant water stress, which are crucial to understanding ecosystem responses to climate change. These examples demonstrate how plants drive and are driven by soil moisture dynamics, and how these interactions affect ecosystem processes and climate conditions.

Quantitative Analysis of Green Water Flows

Understanding bio-geographical patterns of dryland vegetation and the study of plant and microbial response to changes in precipitation conditions require the analysis of soil moisture dynamics and their dependence on soil, climate, and vegetation characteristics. This analysis is complicated because (a) the soil moisture regime is tightly coupled with vegetation dynamics, as plant root uptake is affected by soil moisture fluctuations and one of the major forcings of these fluctuations; (b) the resulting dynamics are strongly nonlinear; and (c) precipitation, a major driver of the soil-water balance, is inherently random. To this end, the study of soil moisture dynamics requires a stochastic representation of the soil-water balance, in which rainfall input is modeled as a random forcing. Thus, the change of soil moisture over time is the result of the difference between water input (rainfall) and output. Rainfall is a sequence of discrete, random events with given average frequency and magnitude, whereas the rates of soil moisture losses from evapotranspiration, drainage, and runoff are nonlinear functions of soil moisture.

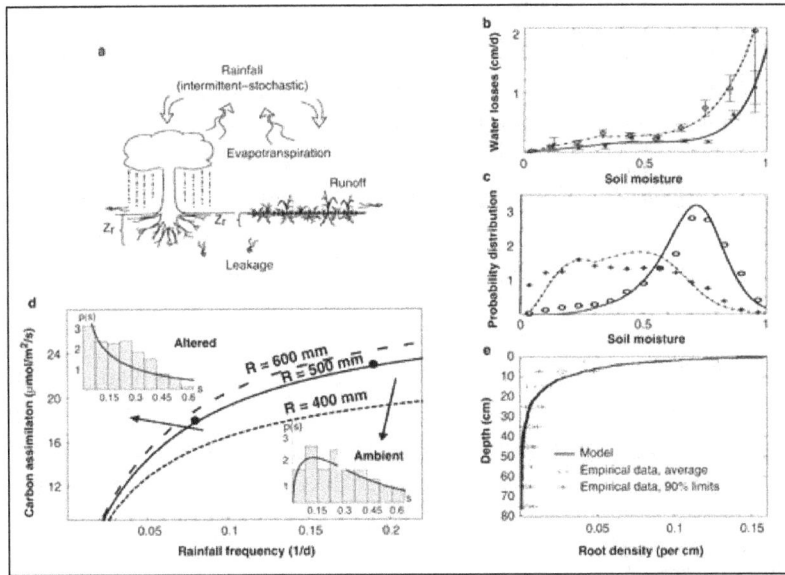

(a) Scheme of the soil moisture balance and its stochastic representation. (b) Soil moisture losses from evapotranspiration and soil drainage, (c) Theoretical and empirical probability distribution of relative soil moisture obtained as solutions of the soil-water balance equation. The lines represent model fits for the March–May (solid line) and June–August (dashed line) seasons to data from Illinois. (d) Dependence of the rate of plant carbon assimilation, An, on rainfall frequency, λ (number of events per day). The theoretical model for different values of annual rainfall, R, is compared with data from a rainfall-exclusion experiment (solid circles and insets) in a mesic grassland. The insets indicate the probability density function of soil moisture. (e) Comparison of theoretical (vertical) root distribution, r(z), from vertically explicit soil moisture model and field data from a shortgrass steppe in Colorado. Abbreviations: cm, centimeters; d, days; m², square meters; μmol, micromoles; R, mean annual rainfall; s, seconds.

Three main classes of stochastic soil moisture models have been proposed in the last decade. A first group of models (known as "bucket models") describes the vertically averaged water balance at the daily time scale within the root zone. This modeling framework does not account for interactions between root-zone soil moisture and water table dynamics and is thus suited for systems with deep water tables. Figure shows the main components of stochastic bucket models of soil moisture. The analysis of the steady-state probability distribution of soil moisture obtained from this model has shed light on the complex and nonlinear interplay among climate, soil, and vegetation, as well as their impacts on the probability of occurrence of plant water stress. This approach provides a more mechanistic means to quantify soil drought conditions than the commonly used hydrological drought indexes. The framework has been also used to investigate the effect of random and intermittent rainfall occurrences on nitrogen cycling, land-atmosphere interactions, and plant carbon assimilation in water-limited ecosystems. The model has also provided a mechanistic explanation of results from rainfall manipulation experiments, including the decline in the rates of carbon assimilation observed in response to a decrease in rainfall frequency (with the total amount of growing-season precipitation remaining the same).

A second class of models includes vertically distributed models, which resolve the vertical profile of soil moisture but do not account for interactions with a shallow water table. These models have been used to relate soil moisture profiles to root structure, providing a theoretical underpinning for the typically observed exponential root profile in relation to climate and soil properties. Figure shows how such vertically distributed models of stochastic soil moisture dynamics can explain the vertical root profile observed in a short-grass steppe. Finally, a third quantitative framework has been developed to investigate the influence of shallow aquifers on soil moisture dynamics in the case of humid environments, where important interactions exist between relatively shallow water tables and root-zone soil moisture. Such models provide quantitative tools to investigate the response of phreatophyte vegetation—or mixed phreatophyte-mesophyte plant communities—to changes in hydrologic conditions.

More sophisticated methods may be needed to capture the spatial heterogeneity in the water content and oxygen availability that characterizes the complex soil environment in cases where microbially mediated processes of mineralization, nitrification, and denitrification need to be taken into account.

Ecohydrological Feedbacks in Terrestrial Ecosystems

Interesting ecohydrological dynamics emerge when there are mutual interactions between the biota and hydrological processes, particularly when they result in a positive feedback. This is the case with species or entire communities that facilitate their own establishment and growth by modifying the rate of hydrological processes. For example, in some wetland forests and marshes, root uptake and transpiration keep the root zone unsaturated, thereby allowing water-intolerant species to survive. Some ecosystems, including tropical montane cloud forests and coastal temperate redwood forests, may rely on canopy-induced inputs of water from canopy condensation and "occult precipitation," that is, the dripping down to the forest floor of moisture stripped by the canopy from fog and low clouds. Similarly, at the regional scale, vegetation may enhance precipitation either through recycling mechanisms or through its effect on roughness, albedo, and soil moisture and associated changes in surface energy fluxes. This vegetation coverinduced enhancement of precipitation may be crucial to plant survival at the desert margins. At the patch scale, dryland soils are often found to be moister under plant canopies than in bare interspaces because of shading, higher infiltration capacity, or hydraulic lift; thus, a nurse plant effect favors seedling establishment under the canopy. In all of these examples, the positive feedback results from plants' ability to enhance their own habitat by affecting the soil water balance.

Positive feedbacks induce bistability in ecosystems. Figure shows the stable and unstable states of vegetation biomass in a model of dryland plant ecosystems accounting for a positive feedback between soil moisture and vegetation. The system is bistable within a range of environmental conditions (i.e. for $R_1 < R < R_2$ in figure). In this case, the dynamics exhibit interesting bifurcations and nonlinear behaviors. Transitions from bistable to stable conditions may occur with abrupt shifts from a stable vegetated state to stable desertlike conditions as R decreases below R_1. Moreover, in the bistability range ($R_1 < R < R_2$), the system has a limited resilience in that it recovers from disturbances (i.e. losses in biomass) only when their magnitude does not exceed a certain threshold Some major ecosystem changes such as deforestation, desertification, or changes in plant community composition have been associated with highly irreversible shifts to alternative, less desirable states. To predict the likelihood of regime shifts to these undesirable states, current research is seeking premonitory signs of state change in the properties of spatial patterns or in the variance of key signals of ecosystem dynamics. However, sound theoretical criteria to recognize symptoms of imminent regime shift in a fluctuating environment are still missing.

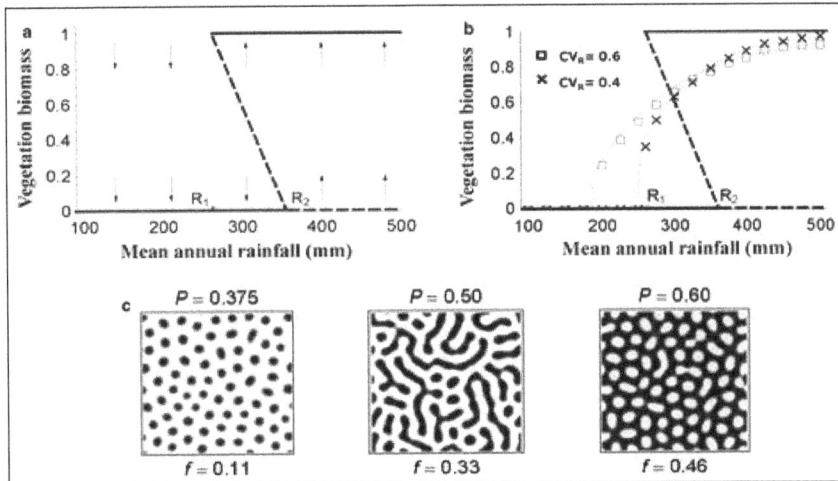

(a) Stable (solid) and unstable (dashed) states of (normalized) vegetation biomass in a dryland ecosystem affected by positive soil moisturevegetation feedbacks. R is the annual rainfall, expressed in millimeters (mm). The arrows indicate convergence toward different stable states. (b) The thick (solid or dashed) lines indicate the deterministic stable and unstable states as in (a). The thin dashed lines represent the stable state induced by random interannual fluctuations of R. As the coefficient of variation of R (CVR) increases, this state exists within a broader range of values of R. Modified from D'Odorico and colleagues. (c) Noise-induced vegetation patterns as a function of the probability, P, of no-water-stress conditions (i.e. P is a proxy for mean precipitation). No patterns emerge with high or low values of P. In intermediate conditions the random alternation of stressed and unstressed states leads to pattern formation with fractional vegetation cover, f, increasing with P.

It has been noticed that these bistable dynamics may be strongly modified by the randomness inherent to hydrologic drivers such as precipitation. In fact, ecosystems that are bistable in the absence of random drivers may be converted into systems with only one stable state as a result of the intensification of random interannual variability of hydrologic fluctuations. Thus, under these conditions the stochastic system is not prone to catastrophic shifts to alternative (undesirable) stable states and is therefore much more resilient than its deterministic counterpart. Overall, noise can play a counterintuitive role in ecohydrological processes: Rather than determining only random fluctuations about the steady states of the underlying deterministic dynamics, these fluctuations can induce new stable states. Environmental variability can also enhance net primary productivity and biodiversity and lead to pattern formation.

Climate change research predicts changes in both the mean and the variance of hydrological drivers. Most of the research on climate change and ecosystems concentrates on the effects of shifts in mean climate and hydrologic conditions. An increase in the variance of environmental fluctuations may have major, complex, and nontrivial effects on population and ecosystem dynamics. Yet the effect of an increase in the amplitude of hydrologic fluctuations remains underappreciated.

Moreover, it is still unclear how ecosystems may respond to the combined effect of hydrologic extremes and climate warming.

Ecohydrology of Blue Water Flows

Below we outline the various implications for ecohydrology of blue water flows.

Ecohydrological Interactions in River Networks

River networks provide important resources and services that are crucial to the functioning of ecosystems and societies. Fluvial networks enhance the ecohydrological connectivity of the landscape and provide preferential pathways for the transport of water, nutrients, sediments, debris, and propagules. Moreover, riparian zones profoundly affect freshwater fish biodiversity, nutrient cycling, food-web dynamics, and energy flows, as well as the spread of pathogens for waterborne diseases such as cholera. Access to blue water resources has historically favored the development of human settlements along river networks and has affected land-use patterns.

Plants growing in the riverbed and in adjacent riparian zones modify the flows of energy, water, and sediment, with important effects on the way riparian corridors and fluvial networks are shaped by hydrological and geomorphic processes. For example, vegetation reduces the energy of water flow shelters sediment beds, consolidates river banks, and enhances sediment deposition. A positive feedback often exists between these abiotic processes and vegetation dynamics. Moreover, biotic processes are affected by interactions between surface water and groundwater, especially in the hyporheic zone, and also by large-scale transport of nutrients through these connected blue water pathways. The understanding of these interactions and feedbacks is crucial to the assessment of nutrient and carbon export from watersheds, and the development of proper strategies for river management and restoration. Altering the flow regime may result in changes in plant community composition in riparian ecosystems, as different plant species are adapted to different hydrologic conditions. For example, the effectiveness of waterborne dispersal (hydrochory) is both species and flow dependent. In fact, the hydrochory of nonbuoyant propagules requires high-flood events, whereas low-density floating seeds can be transported long distances even in low-flow and low-energy streams. Moreover, the timing of high- and low-flow seasons affects the success of seed germination and seedling establishment.

Ecohydrological Significance of River Network Structure

Recent studies have provided new quantitative frameworks explaining how the complex topology of river networks affects ecohydrological transport and ecological patterns. For example, the speed of migrating population fronts has been related to topological features of the network structure. This structure also affects the spatial dynamics of the biota in riparian corridors, as it imposes an anisotropic configuration in the set of possible connections and interactions among different landscape units, and it also controls the dominant patterns underlying the spread of species, pathogens, or other agents along the waterways.

Biodiversity patterns of fish populations in large river basins have been related to the dendritic structure of the river network through neutral metacommunity models, whereby local communities are connected through links imposed by the geomorphology of the channel network.

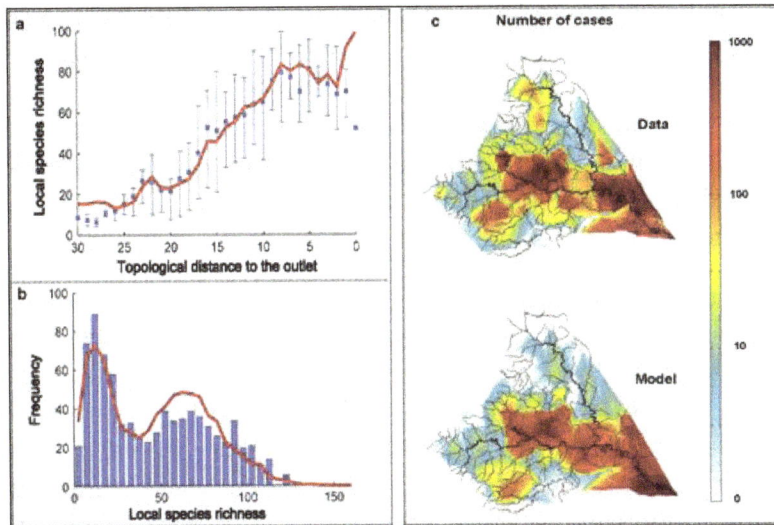

(ab) Patterns of local species richness (LSR) of freshwater fish in the Mississippi-Missouri basin. The location relative to the network structure is a determinant factor of fish diversity patterns. The LSR profile displays the average α-diversity of all sites located at the same distance from the outlet throughout the network. The observed and simulated LSR profiles show a clear increase in the downstream direction resulting from the converging character of the river network (a). The bimodality in the frequency distribution of LSR (b) reflects the difference in biodiversity between the western and eastern parts of the basin. (c) Hydrologically controlled spatial distribution of cumulated cholera cases in two years following an epidemic outbreak (Tukhela River basin, South Africa).

The spread of waterborne diseases such as cholera is also controlled by river network structure. Figure illustrates observed and computed cumulated cholera cases within a fluvial domain using a spatiotemporal model of disease epidemics where local communities are reached by pathogens moving along the fluvial network. An emerging line of research in ecohydrology is concerned with the effect of hydrologic controls on the spread of water-related diseases such as cholera, malaria, dengue hemorrhagic fever, shigellosis, and schistosomiasis (or bilharzia). These controls may determine seasonality in the recursion of disease or the susceptibility of a community to infections and possibly pandemics, for instance by affecting the local thresholds for epidemic outbreaks.

Isotopes and Hydrology

"Isotope Hydrology" is a relatively young scientific discipline (or rather an interdisciplinary field), when it was first realised the methods of nuclear physics for the detection of isotopes could have useful applications in hydrology. The classical tools of isotope hydrology are the isotopes of the constituents of the water molecule (H_2O) itself, namely the rare stable isotopes of hydrogen and oxygen (^2H, ^{18}O) and the radioactive tritium (^3H). These were soon complemented by radiocarbon (^{14}C), which enabled water dating via the decay of ^{14}C in the carbon dissolved in the water.

Later on, many more methods were added to the toolbox of the isotope hydrologists, some of which are not really isotope methods at all. For example, dissolved gases such as noble gases and certain anthropogenic trace gases are very useful markers to "trace" the fate of water in some parts of the hydrological cycle. Such substances are thus called tracers, or more precisely "environmental tracers" to distinguish from substances that are added purposefully to the water (e.g. dyes). From a modern point of view, isotope hydrology may therefore be defined as the application of „environmental isotopes and tracers to study (parts of) the hydrological cycle.

Isotope hydrology is a truly interdisciplinary science. It emerged from the application of methods developed in physics (analytical techniques) to problems of Earth and environmental sciences. This combination proved very successful not only in hydrology but also other branches of Earth sciences, leading to the related fields of isotope geology and isotope geochemistry. Hydrology, the science of water, on the other hand, is in itself a very diverse and segmented field. Some parts are covered by engineers (hydraulics, water supply and waste water), others by geologists (groundwater: hydrogeology), and still other special disciplines such as oceanography (water in the oceans), meteorology (water in the atmosphere), and limnology (water in lakes and rivers). Isotope methods have become important in physical oceanography and limnology, but the term isotope hydrology is most often used in relation to the study of groundwaters.

A central role in the development of isotope hydrology and its worldwide application has been played by the International Atomic Energy Agency (IAEA) in Vienna. This at first sight maybe surprising link grew out of the monitoring of radioactive fallout from the bomb tests, which added large quantities of tritium to the water cycle. This effort resulted in the establishment of the "Global Network of Isotopes in Precipitation" (GNIP), which is now run jointly by the IAEA and the WMO (World Meteorological Organisation). GNIP-stations measure stable isotopes and tritium in precipitation, producing an invaluable basic data set for the science of isotope hydrology. The IAEA runs itself an isotope lab, which distributes international isotope standards to the research laboratories worldwide. The IAEA also organises regular conferences on isotope hydrology since 1963.

Problems studied in isotope hydrology may be classified in two large clusters: Determination of the origin of water masses and the conditions during formation:

- Identification and separation of water components.

- Determination of groundwater recharge areas, flow paths, mixing.

- Determination of the origin of contaminants.

- Reconstruction of recharge temperatures for palaeoclimate studies.

Determination of the residence time of water in the system ("water age"):

- Calculation of flow velocities, assessing mixing and dispersion.

- Determination of water fluxes, recharge rates, and exchange rates.

- Study of transport and degradation of contaminants.

Environmental Tracers

A tracer is a substance, which is present only in trace concentrations ("Spurenstoff") and marks a trace in a natural system (marker). In hydrology, it ideally it marks the water itself or dissolved substances therein, and moves passively with the water or the solutes.

Isotopes are often nearly ideal tracers, e.g:

- Isotopes of H and O mark the water molecule.

- Isotopes of C, N, S etc. Mark dissolved substances.

- Conservative solutes (e.g. Noble gases) are good tracers.

Environmental tracers are isotopes or other tracers that are widely distributed in the environment, especially the hydrosphere. They can be both of natural or anthropogenic origin. They can be further classified into:

- Transient tracers: Anthropogenic substances with an input to the environment (hydrosphere) that varies with time. Typical examples are tritium, CFCs, SF_6, and ^{85}Kr.

- Geochemical tracers: Tracers of natural origin with a (more or less) constant input, e.g. stable isotopes, ^{14}C, noble gases.

The term "environmental tracers" is mainly used to distinguish them from "artificial tracers", both of which are often simply called tracers but are used in quite different settings and by largely different researchers. Artificial tracers are released deliberately into the investigated system, in local tracer experiments (e.g. dyes, chemicals, also isotopes, SF_6 and so on). The essential difference is that environmental tracers can be applied for studies on large spatial and temporal scales, whereas artificial are suitable for targeted studies of small scale processes.

Isotopes

Nuclei of atoms consist of protons and neutrons (= nucleons):

- Z = number of protons (atomic number)

- N = number of neutrons

- A = Z + N = number of nucleons (mass number).

Z determines to which element a nucleus belongs and is equal to the number of electrons in a neutral atom. The number of electrons determines the chemical properties of the elements.

As a general term for the type of nucleus determined by Z and N we use the word nuclide. Isotopes are nuclides of the same element (same Z), with a different number of neutrons (thus varying N and A). The term is derived from the Greek iso topos = same place (in the periodic system).

There are two forms of notation for isotopes:

Full notation: $_Z^A X_N$, e.g. $_6^{14}C_8$

Short notation: $^{A}X, e.g.\ \ ^{14}C$

The full notation is rarely used, because the short notation contains all information, if one knows the number of protons (Z) of the elements.

The Chart of the Nuclides

The chart of the nuclide chart provides an overview of the existing isotopes and is an important source of basic information for isotope sciences. The isotopes are mapped in plot of neutron number N versus proton number Z. Isotopes of a single element stand in rows. Stable isotopes of the light elements align roughly along the 1:1 diagonal, for heavier elements, there is always an excess of neutrons over protons.

The chart contains usually a lot of information on the nuclides, such as exact mass or half-life and decay mode for the radioactive isotopes. Nuclide charts and the related information can be found on the web, e.g. at the Korea Atomic Energy Research Institute.

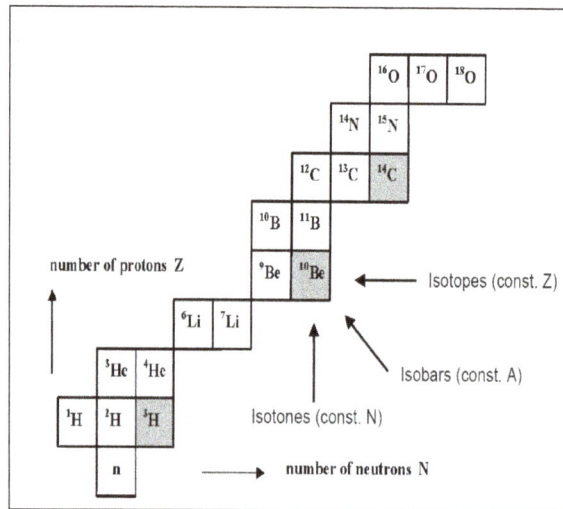

The lowest part of the chart of the nuclides

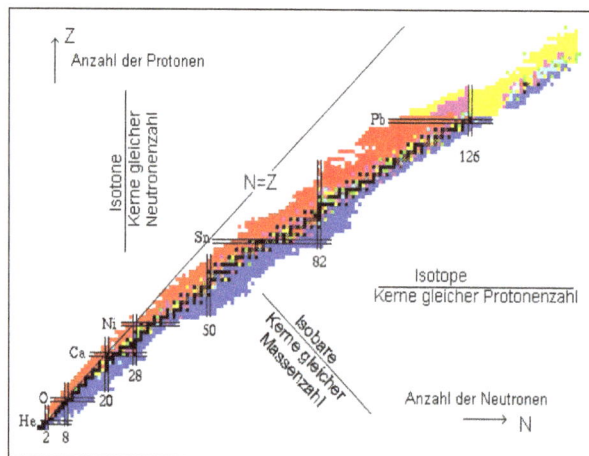

The entire chart of the nuclides. Black squares correspond to stable isotopes, all coloured squares to radioactive isotopes. Colours indicate different decay modes. The highlighted proton and neutron numbers are the so-called "magic numbers".

Stability and Abundance of Stable Isotopes

The most important distinction for applications of isotopes is between stable and unstable (radioactive) isotopes. Stability of a nucleus requires a balance between the electromagnetic force leading to a repulsion between the protons and the strong force that brings about an attraction between the nucleons. This is the reason why a sufficient number of neutrons is always necessary to "glue" the protons together.

Some rules of thumb on stability:

- For the light elements, nuclides with $Z = N$ (e.g. 12C, 14N, 16O) or a slight neutron excess (e.g. 13C, 15N, 18O) are stable.

- For the heavy elements, only nuclides with a strong neutron excess are stable (e.g. 208Pb, $Z = 82$, $N = 126$).

The abundance of the isotopes is mainly determined by processes during nucleosynthesis, which took place during the big bang and in supernovae explosions. A high chance of stability occurs if Z and/or N equal the so-called "magic numbers", i.e. 2, 8, 20, 28, 50, 82, 126. Particularly stable are the doubly magic nuclei, where both Z and N are magic numbers. Due to their high stability, such nuclides have a large natural abundance (e.g. ^{4}He (2,2), ^{16}O (8,8), ^{40}Ca (20,20), and ^{208}Pb (82,126)). On the other hand, nuclides with uneven Z tend to be less stable, and isotopes with both uneven Z and uneven N are usually rare.

For applications in environmental science, the stable isotopes of light elements are particu- larly important. As expected, symmetric ($Z = N$) and (doubly) magic nuclides are most abundant (e.g. ^{4}He, ^{12}C, ^{14}N, ^{16}O, ^{20}Ne), whereas isotopes with uneven Z or N are rather rare (e.g. ^{3}He, ^{13}C, ^{15}N, ^{17}O).

Table: Isotopic abundance (ratio of the abundance of a given isotope to that of all isotopes of the element) of the stable isotopes of the light elements. There is always one dominating light isotope, and one or two rare, heavier isotope(s) with higher neutron number.

Element	Abundance of stable isotopes (%)		
Hydrogen	1H 99.985	^{2}H 0.015	
Helium	4He 99.9999	^{3}He 0.0001	
Carbon	^{12}C 98.89	^{13}C 1.11	
Nitrogen	^{14}N 99.63	^{15}N 0.37	
Oxygen	^{16}O 99.758	^{18}O 0.204	^{17}O 0.038
Neon	^{20}Ne 90.51	^{22}Ne 9.22	^{21}Ne 0.27

Radioisotopes and their Production

Radioactivity

Most nuclei are actually unstable and undergo radioactive decay. There are many very short- lived isotopes, which have no practical importance, but some longer-lived isotopes are of great use in

Earth and environmental science and especially in isotope hydrology (e.g. ³H, 14C). We therefore look a bit closer at the phenomenon of radioactivity.

There are several different modes of radioactive decay. Particularly important for the isotopes that are applied in hydrology are the α and β^- decay. In an α-decay an α-particle, which is nothing else than the very stable ⁴He nucleus, is emitted. The proton and neutron numbers of the decaying nuclide are thus both reduced by 2, and the mass number is reduced by 4. In a β-decay a β-particle, which is nothing else than an electron, is emitted, and a neutron is converted to a proton. Consequently, the proton number is enlarged by 1 while the neutron number is reduced by 1, and the mass number remains the same. The effects of radioactive decays on the position of an isotope in the chart of nuclides is shown in figure.

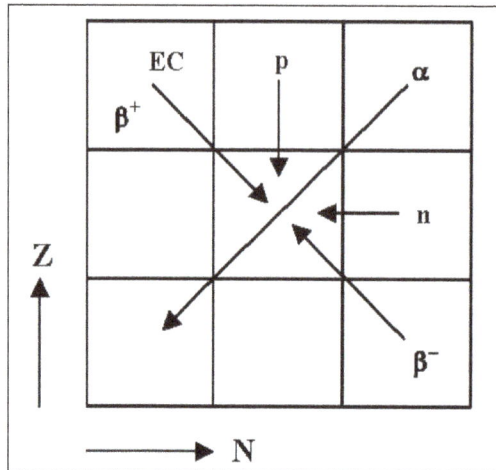

Effects of radioactive decay modes in the chart of the nuclides

β^- decay: $n \rightarrow p + e^- + \bar{\nu}_e$ $^A_Z X_N \rightarrow ^A_{Z+1} Y_{N-1}$

β^+ decay: $p \rightarrow n + e^+ + \nu_e$ $^A_Z X_N \rightarrow ^A_{Z-1} Y_{N-1}$

e⁻ capture: $p + e^- \rightarrow n + \nu$ $^A_Z X_N \rightarrow ^A_{Z-1} Y_{N+1}$

α decay: $^A_Z X_N \rightarrow ^{A-4}_{Z-2} Y_{N-2} + \alpha$ $\alpha = ^4_2 He_2$

fission: $X \rightarrow$ 2 daughter nuclides + some n

γ decay: $^A_Z X_N^* \rightarrow ^A_Z X_N + \gamma$ (no change of nuclide)

Radioactive decay modes

Radioactive decay is a purely statistical process, with constant decay probability. The number of decays in a given time interval is proportional to the total number of nuclides:

$$\frac{dN}{dt} = -\lambda N$$

with λ [T^{-1}] is the decay constant, i.e. the decay probability per unit time. The decay constants of the different radioisotopes are characteristic material constants. They are in principle well-known, although the precision of the values for long-lived isotopes is not always very high.

Integration of equation $\frac{dN}{dt} = -\lambda N$ with the initial condition $N(0) = N_0$ yields:

$$N(t) = N_0 \cdot e^{-\lambda t}$$

This is the well-known equation of an exponential decay. The time dependence of this decay is entirely characterised by the decay constant. Instead of λ, it may be more convenient to use its inverse $\tau = 1/\lambda$, which has the dimension of a time. λ corresponds to the time when the number of nuclides has declined by a factor of $1/e$ and it equals the expectation value for the lifetime of the nuclides. It is thus called the mean life (mittlere Lebensdauer) of the isotopes. Although mathematically less elegant, a much more customary measure of the characteristic time of a radioactive decay process is the half-life (Halbwertszeit) $T_{1/2}$, which is the time when the number of nuclides has declined to half of its initial value. It can easily be seen that the half-life is linked to t by a factor of ln2:

$$T_{1/2} = \frac{In\,2}{\lambda} = \tau \cdot In\,2 \quad \left(\text{follow from } N\left(T_{1/2}\right) = N_0 \cdot e^{-\lambda T_{1/2}} = \frac{N_0}{2} \right)$$

Equation above is a simple but very useful relationship. It may even be worthwhile to remember that ln2 is about 0.7, thus the half-life is about 70% of the mean life.

(A related useful rule of thumb is that the doubling time of the money in your bank account is 70 yr divided by the interest rate in %. The interest rate corresponds to λ, if expressed in % we have to multiply ln2 by 100, leading to the 70 yr constant. The same is true for the doubling time of population and the yearly growth rate, e.g. the population of Egypt with a current growth rate of 2 % per year will double in 35 yr if nothing changes).

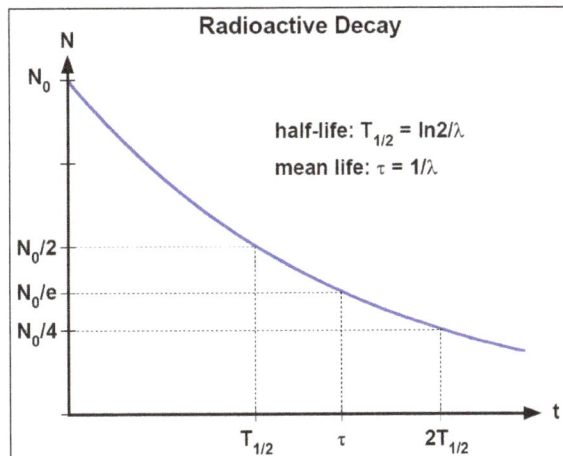

Radioactive decay curve, and definition of some characteristic times

Radioactive isotopes can yield time information (age), if N(t) and N_0 are known (the λ's are known constants). The age is obtained simply by solving equation $N(t) = N_0 \cdot e^{-\lambda t}$ for the time:

$$t = -\frac{1}{\lambda} In\left(\frac{N(t)}{N_0}\right) = \tau In\left(\frac{N_0}{N(t)}\right)$$

In applications of this equation for dating, time zero is some unknown date in the past, whereas time t is usually the present time or more precisely the time of analysis. N(t) is thus directly measured, the problem is to know N_0. The basic idea is usually that N_0 is assumed to be a known constant value. Another very important approach is to measure the product of the decay (the so-called daughter isotope) in addition to the decaying radioisotope itself (the mother isotope).

Radioisotopes are usually rather rare (because they constantly disappear), but the fact that they decay offers an opportunity to detect them even at very low abundance. Thus, in practice, often not N (number of atoms, abundance) is measured, but the activity, which is the number of decays per unit time:

Definition of activity:

$$A \equiv \frac{dN}{dt} = -\lambda N$$

Units for A are:

- Becquerel (Bq) = s^{-1} (decays per sec).

- Curie (Ci): 1 Ci = $3.7 \cdot 10^{10}$ Bq (activity of 1 g Ra).

The modern SI-unit is of course the Bq, which we will use. The Curie is a historical unit, de- rived from the early experiments of Marie Curie with radium as a source of radioactivity. The Curie is a rather large unit environmental activities are often in the range of pCi. It should be noted that activity is a measure of the mass (number of atoms) of a radioactive substance. Their concentrations are often given as activity concentrations (e.g. Bq/m^3).

Radioisotopes, their Abundance and Production

Radioactive isotopes can only occur in nature if they are either very long lived (i.e. the half- life is comparable to the age of Earth) or if they are constantly produced by nuclear processes. Radioisotopes can be classified according to their origin:

- Primordial (remnants from the formation of the solar system, when supernova- produced radioisotopes must have been present).

- Cosmogenic (produced by interactions with cosmic rays in the Earth atmosphere or near the Earth's surface).

- Subsurface-produced (produced by nuclear reactions in the solid Earth).

- Anthropogenic (produced by technical processes).

Most radioisotopes have very low isotopic abundances, except for some very long-lived primordial isotopes. Some heavy elements occur in nature, although they have no stable isotopes: Uranium, thorium, and isotopes produced in the U/Th-decay series.

Radioisotopes can be produced by nuclear reactions induced by irradiation of nuclides. Sources and types of irradiation are:

- In the atmosphere, near the Earth's surface: Cosmic rays (CR, protons, neutrons, etc.).

- In the subsurface: U/Th-decay series and fission (α s and neutrons).

- In the anthroposphere: Nuclear bombs, reactors, accelerators, etc. (mainly neutrons).

Table: Isotopic abundance, half-life, and decay mode (EC: electron capture, SF: spontaneous fission) of long-lived primordial radioisotopes. The half-life of ^{238}U is quite close to the age of Earth (about 4.55 Gyr), some others are much longer.

Isotope	Abundance (*)	Half Life (yr)	Decay Mode
^{40}K	0.0117	1.28×10^9	β^-, EC
^{50}V	0.250	1.4×10^{17}	EC, β^-
^{87}Rb	27.83	4.75×10^{10}	β^-
^{144}Nd	23.8	2.29×10^{15}	α
^{148}Sm	11.3	8×10^{15}	α
^{176}Lu	2.59	4.0×10^{10}	β^-
^{174}Hf	0.162	2.0×10^{15}	α
^{232}Th	100	1.45×10^{10}	α, SF
^{235}U	0.72	7.038×10^8	α, SF, Ne
^{238}U	99.27	4.468×10^9	α, SF

An example of a nuclear reaction that produces radioisotopes is spallation, where at high- energy particles of the cosmic radiation break a nucleus apart, forming a radioisotope and some smaller fractions, e.g:

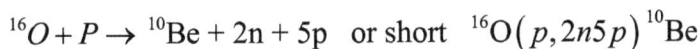

$$^{16}O + P \rightarrow\ ^{10}Be + 2n + 5p \quad \text{or short} \quad ^{16}O\left(p, 2n5p\right)\ ^{10}Be$$

Another example is thermal neutron capture, where low-energy neutrons react less violently with nuclides, e.g.:

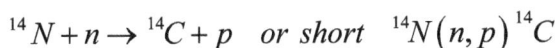

$$^{14}N + n \rightarrow\ ^{14}C + p \quad \text{or short} \quad ^{14}N\left(n, p\right)\ ^{14}C$$

Which isotopes are produced in the atmosphere (by cosmic rays) or the subsurface (by particles from the U/Th chains) depends mainly on the nuclides that are available for reaction. Main target elements are:

- In the atmosphere: N, O, Ar

- In rocks: Li, O, Na, Mg, Al, Si, Cl, K, Ca, Fe.

Table: Examples of nuclear reactions producing radioisotopes in the atmosphere and in the subsurface.

Atmospher	Subsurface
^{14}N (n, p) ^{14}C	^{6}Li (n, α) ^{3}H
^{14}N (n, ^{3}H)^{12}C	^{35}Cl (n, γ)^{36}Cl
^{40}Ar (p, 3n2p)^{36}Cl	^{40}Ca (n,2n 3p) ^{36}Cl
^{40}Ar (n, 2n)^{39}Ar	^{36}K (n, p)^{39}Ar
^{40}Ar (p, sp)^{26}Al	^{40}Ca (n, α)^{37}Ar

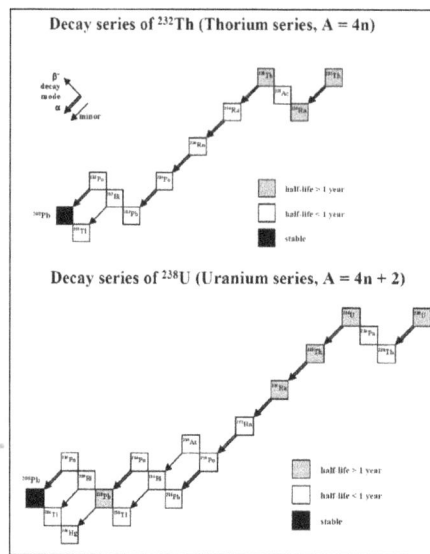

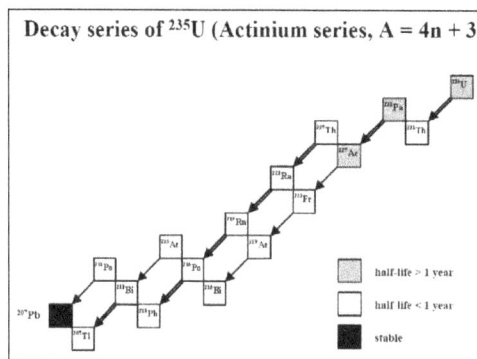

Radioactive decay series of uranium and thorium

Very important for environmental isotope science is the production of radioisotopes by cosmic radiation. Cosmic rays (CR) are divided in two types:

- Primary CR: 87 % protons (p), 12 % α-particles, 1 % heavier nuclei.

- Secondary CR: mainly neutrons.

Primary cosmic rays are important in the upper atmosphere, below the secondary radiation dominates. Figure shows the cascades of cosmic rays occurring in the atmosphere.

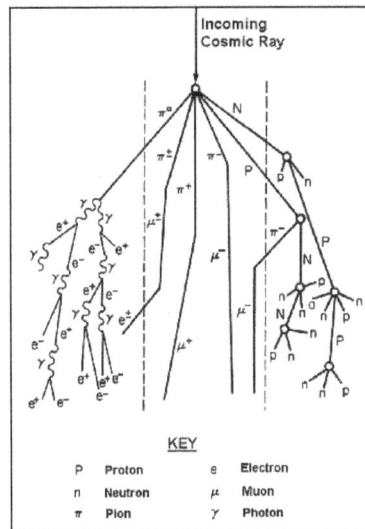

Cosmic ray shower in the atmosphere, by reaction with nuclei of gases in the air

Production rates by cosmic radiation are very low, hence cosmogenic isotopes are very rare, e.g:

- $^{14}C/^{12}C \sim 10^{-11}$

- $^{36}Cl/^{35}Cl \sim 10^{-15}$

- $^{3}H/^{1}H \sim 10^{-18}$, global natural ^{3}H inventory: 3.6 kg

Table: Production rates of some cosmogenic isotopes of environmental importance.

Radioisotope	Prod. Rate (atoms cm^{-2} s^{-1})
^{3}H	0.28
^{7}Be	0.035
^{10}Be	0.0018
^{14}C	2.18
^{36}CI	0.0019

Another important source of radionuclides is anthropogenic production. Major sources of anthropogenic radionuclides are:

- Tests of nuclear weapons in the atmosphere.

- Accidents of nuclear power plants.

- Normal releases of nuclear power and reprocessing plants.

- Waste from other applications (medical, science, etc).

Anthropogenic nuclides of importance for isotope hydrology are listed in table. The primary example of such nuclides is of course tritium, which along with the stable isotopes of H and O is the classical tool of isotope hydrology. The input history of tritium into the hydrosphere via precipitation is illustrated by figure. showing the prominent peak resulting from atmospheric nuclear bomb testing, the so-called "bomb peak" that occurred in 1963.

Table: Anthropogenic nuclides of importance for isotope hydrology.

Isotope	Half-life (yr)	Origin
^{3}H	12.32	Bomb tests
^{14}C	5730	Bomb tests
^{36}CI	308'000	Bomb tests
^{85}Kr	10.7	Fuel reprocessing

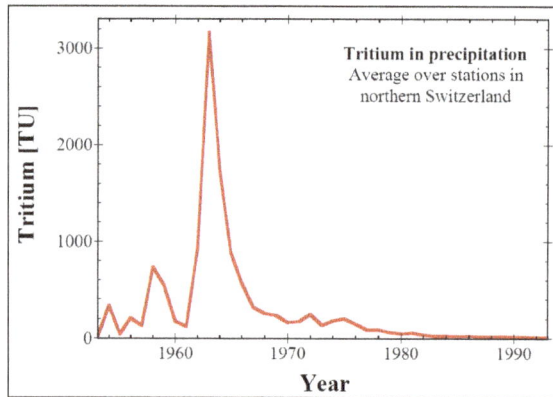

Tritium in precipitation of northern Switzerland. The shape of the curve with the pronounced "bomb peak" in 1963 is typical for the entire northern hemisphere.

Isotopes and Tracers in Hydrology

Table summarises the isotopes that are applied in hydrology and related fields. The stable isotopes deuterium and ^{18}O as well as the radioactive tritium are direct markers of the water molecule. All other isotopes are related to some dissolved substance, such as dissolved inorganic/organic carbon (DIC, DOC), sulphate (SO_4), or dissolved (noble) gases.

The half-lives of the radioisotopes cover a wide range from the short-lived ^{222}Rn (3.8 days) to the long-lived ^{81}Kr (210 kyr) and ^{36}Cl (308 kyr). Consequently a wide range of ages is at least in principle accessible for dating by radioisotopes in water. Dating is not only possible through the observation of the radioactive decay, but also by using the time-dependent input of some radioisotopes or the steady accumulation of some stable isotopes. A special case among the dating methods is the mother-daughter pair ^{3}H-^{3}He.

In addition to dating, many (stable) isotopes can be used as markers and to obtain information about the conditions during the formation of a water mass.

Table: Isotopes that are used in hydrology.

Isotope	Phase	Half Life	Use
Deuterium (^{2}H, D)	H_2O (HDO)	Stable	Marker, formation

Tritium (^{3}H, T)	H$_2$O (HTO)	12.32 yr	Dating (input)
Heiium-3 (^{3}He)	Dissolved He	Stable	Dating (^{3}H/^{3}He)
Heiium-4 (^{4}He)	Dissolved He	Stable	Dating (accumulation)
Carbon-13 (^{13}C)	DIC, DOC,...	Stable	Marker
Carbon-14 (^{14}C)	DIC, DOC,...	5730 yr	Dating (decay)
Nitrogen-15 (^{15}N)	Diss. N$_2$, NO$_3$,...	Stable	Marker
Oxygen-18 (^{18}O)	H$_2$O (H$_2$^{18}O)	Stable	Marker, formation
Suifur-34 (^{34}S)	SO$_4$...	Stable	Marker
Chlorine-36 (^{36}CI)	CI$^-$	308'000 yr	Dating (decay)
Chlorine-37 (^{37}CI)	Chcs, ...	Stable	Marker
Argon-39 (^{39}ar)	Dissolved ar	269 yr	Dating (decay)
Argon-40 (^{40}ar)	Dissolved ar	Stable	Dating (accumulation)
Krypton-81 (^{81}kr)	Dissolved kr	210'000 yr	Dating (decay)
Krypton-85 (^{85}kr)	Dissolved kr	10.7 yr	Dating (input)
Radon-222 (^{222}rn)	Dissolved rn	3.8 d	Dating (accumulation)
U-series (^{238}U, ^{234}U,..)	Dissolved UO$_2$	Variable	Dating (disequilibrium)

Another category of environmental tracers in hydrology are dissolved conservative gases, especially the noble gases He through Xe, which in particular can be used to determine the temperature that prevailed during the infiltration of groundwater. The transient gas tracers CFCs (chlorofluorocarbons, Freons) and SF$_6$ (sulfurhexafluoride) are used for dating young groundwater.

Table: Gases and other tracers that are used in hydrology.

Gases			
Tracer	Phase	Half Life	Use
He, Ne, Ar, Kr, Xe	dissolved	stable	Formation (temp.)
Nitrogen (N^2)	dissolved	stable	Formation (temp.)
CFCs (11, 12, 113)	dissolved	stable (oxic)	Dating (input)
SF$_6$	dissolved	stable	Dating (input)
Others			
Tracer	Phase	Half Life	Use
Temperature	bulk	stable	Marker
Conductivity (TDS)	bulk	stable	Marker
Ions (CI$^-$, Li$^+$,....)	dissolved	stable	Marker

A final category of environmental tracers are such simple properties or constituents of the water as temperature and dissolved ions. Dissolved ions can be looked at individually or as a sum parameter such as salinity or TDS (total dissolved solids), which often is parameterised through the more easily measurable electrical conductivity. These properties, or at least temperature and conductivity, should always be measured in applications of tracer methods in aquatic systems. Temperature and conductivity enable the calculation of density and thus form the basis of any assessment of the physical state of the system. Moreover, they can very often be used as quasi-conservative tracers that are very simple, quick and cheap to measure. When more sophisticated tracer methods are applied, one should not forget that sometimes a lot of useful information can already be gained from these simple tools.

Dating Methods and Ranges

Among the many applications of environmental tracers in hydrology, water age dating, i.e. the determination of the residence time of a water mass in a particular system is probably the most important. There is a large variety of dating tracers. These can be classified in four categories with respect to the basic principle of the dating methods:

- Radioactive decay: Direct use of the decay law with the difficulty of knowing the initial (activity) concentration C_0. The classical example is ^{14}C, but ^{36}Cl, ^{39}Ar, and ^{81}Kr also belong to this category. ^3H actually usually works via input variation.

- Mother-daughter pair: Combined determination of the decaying radioisotope and its accumulating stable daughter, solving the problem of C_0 (which equals the constant sum of mother and daughter). The only relevant example is ^3H-^3He.

- Accumulation: Observation of only the accumulation of a stable radiogenic daughter isotope, with the problem of knowing the production and accumulation rate.

- Examples are ^4He and ^{40}Ar, where the mother isotopes (U/Th and ^{40}K) are so long-lived, that accumulation can assumed to be linear with time.

- Input variation: Using the time information in the variable input history of transient anthropogenic tracers. Examples are ^3H, ^{85}Kr, CFCs, and SF_6.

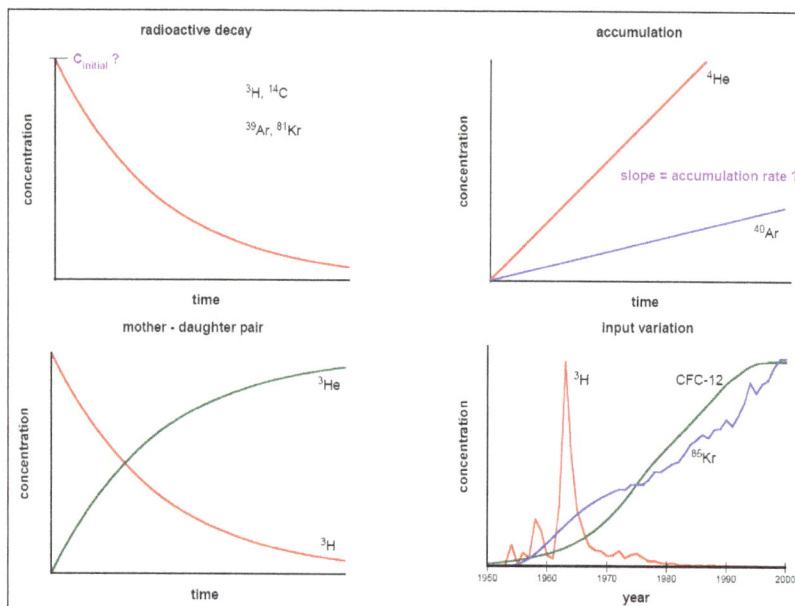

The four basic principles of the dating methods in hydrology

The dating methods cover a wide dating range. Obviously, the temporal range of methods based on radioactive isotopes are determined by the half-life of the isotope in question. As a rule of thumb, the dating range extends at most up to 10 half-lives, when is starts to become very difficult to detect the small remaining concentrations. The transient tracers are also clearly limited to the period when their input variation took place, which is essentially the last 50 years. This limit also applies for ^{3}H and the ^{3}H-^{3}He method. The accumulation of stable daughters of very long-lived mother isotopes in principle allows dating on very long time scales. In practice, however, the 4He and ^{40}Ar accumulation rarely work as quantitative dating tools, because the accumulation rates are not known sufficiently well.

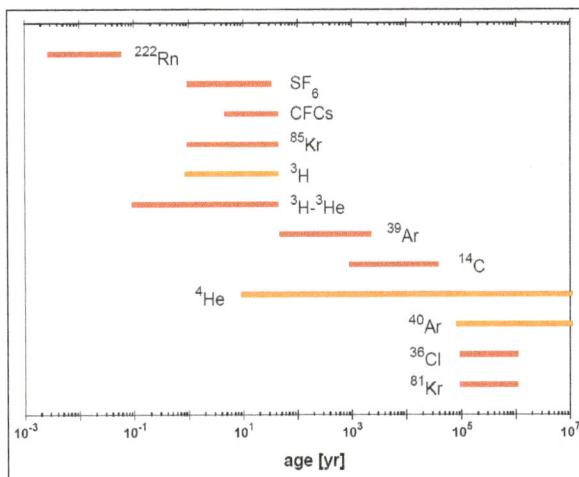

Dating ranges of the different tracer methods. Quantitative methods are indicated in red, whereas orange bars refer to methods that usually give only qualitative age information

Water, the Hydrological Cycle and Basics of Hydrology

Hydrology is of course the science of water, but was does this mean exactly. Water is an ubiquitous substance on Earth, it is present in all spheres – not only the hydrosphere, but from the lithosphere to the atmosphere and the biosphere to the anthroposphere. So what exactly is hydrology concerned with.

Classical hydrology may be defined a bit more precisely as the science of water on the conti- nents, above, on, and below the surface. This especially means that it does not include the ocean, which is treated by the field of oceanography. However, since this lecture treats the physics of aquatic systems, we explicitly include the oceans (although they receive relatively little attention). In fact, tracer oceanography is a large field that is of course closely linked to tracer and isotope hydrology, which was chosen as subtitle of this lecture. But a complete coverage of tracer oceanography would require another, dedicated lecture.

Hydrology is a physical science, it deals with such issues as water (mass) balance and water flux- es, or generally the movement of water in space and time. The classical hydrological systems are catchments of rivers and aquifers, where quantities such as precipitation, recharge, storage, and discharge are determined. Hydrology is to a large extent focussing on quantitative aspects, such as the study of river discharge (e.g. for flood prediction).

However, it also has chemical, biological, and technical/sociological aspects. Some related aspects are dealt with by the engineering sciences, e.g. water supply, waste water, flood protection, etc.

Sub disciplines of hydrology or disciplines related to it are:

- Hydrogeology and soil physics (water in the saturated and unsaturated zones).

- Physical Oceanography and Limnology (water in the ocean and lakes/rivers).

- Meteorology and (palaeo)climatology (water in the atmosphere, precipitation).

The hydrosphere includes all forms of water on and below the surface of the Earth. This includes the ocean and all parts of the global water cycle except the atmospheric part (water vapour, precipitation). The global hydrological cycle is the framework in which the different aquatic systems exist, that we will study. Physics of aquatic systems, as defined in the first part of this lecture cycle, and in contrast to hydrology, looks not so much at the flow of water in the cycle and between the subsystems, but rather at the internal physical processes in the single subsystems (e.g. density stratification, currents, transport, mixing, etc. in the ocean, in lakes, and in aquifers).

Table lists the compartments of the hydrosphere. 70 % of the Earth's surface are covered by oceans, which contain 97 % of the water in the hydrosphere (including the cryosphere). The largest reservoir of fresh water are the polar ice sheets, followed by groundwater. Lakes are a comparatively minor reservoir. Of course, due to their accessibility, lakes and rivers are important fresh water resources from the point of view of human use, although in many parts of the world groundwater is the dominant source of drinking water. The mean hydrologic residence time (= volume/throughput) of the ocean water is about 3000 years, the one of the atmospheric water vapour only some 10 days. On the continents the water spends on average some 300 years, but with enormous differences between different reservoirs (short residence times in rivers and lakes, long residence times in ice and groundwater).

Table: Compartments of the hydrosphere.

	Volume in 10^3 km³	% of total freshwater	flux 10^3 km³/year	turn-over time year
Salt Water				
Oceans	1 350 000		425	3000 [1])
Freshwater				
Ice	27 800	69.3	2.4	12 000 [2])
Groundwater	8 000*	29.9	15	500 [3])
Lakes	220**	0.55		
Soil moisture	70	0.18	90	0.8 [4])
Atmosphere	15.5	0.038	496	0.03 [5])
Reservoirs	5	0.013		
Rivers	2	0.005	40	0.05 [6])
Biomass	2	0.005		
Total	40 114	100		

Figure shows a very crude scheme of the global hydrological cycle, with the fluxes between the major compartments. The annual evaporation from the oceans amounts to 425'000 km³ of water, most of which returns directly as precipitation onto the oceans. 40'000 km³ of water are "exported" to the continents and return as river discharge. The precipitation of 111'000 km³ of water per year onto the continents is thus to nearly 2/3 "home-made" (evaporation from the continents), only 1/3 is imported from the ocean. These 111'000 km³ of water per year are in principle the renewable resource of fresh water, however only a fraction of this water, which has been estimated to 12'500 km³/yr, is really accessible.

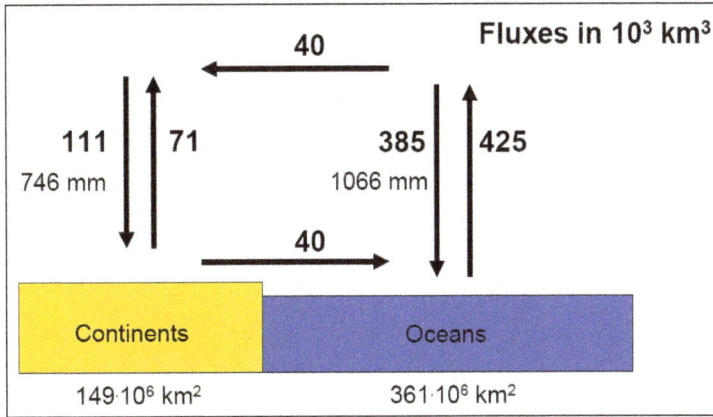

Crude scheme of reservoirs and fluxes in the global hydrological cycle. Bold numbers indicate fluxes in units of 10^3 km^3/yr. The yearly precipitation fluxes are also given in mm, calculated with respect to the surface areas of continents and oceans, respectively. The total global evaporation/precipitation flux with respect to the Earth's surface amounts to very nearly 1 m per year.

A more detailed picture of the hydrological cycle is given by figure. The processes of interest for the hydrological water balance are precipitation, evaporation, evapotranspiration, surface discharge, and groundwater recharge (infiltration). Lakes and aquifers act as temporary storage reservoirs. While for the hydrological balance only the mean residence time of the water in these reservoirs is relevant, we will also be concerned with their internal structure and the internal transport processes.

The distribution of precipitation on Earth is very uneven, as shown by Figure. Tropical forests and coastal mountain ranges receive most precipitation. The subtropics and the inner regions of the large continents of the northern hemisphere are dry. Water scarcity prevails at the places where dry climate coincides with high (and often still increasing) population density (e.g. northern Africa, the Middle East, and northern China). Most of the water consumption goes into irrigation of fields to produce food. The worldwide demand for irrigation water is about 3'00 km^3/yr, or nearly one fourth of the available renewable resource. This shows clearly that the human interference with the water cycle is substantial.

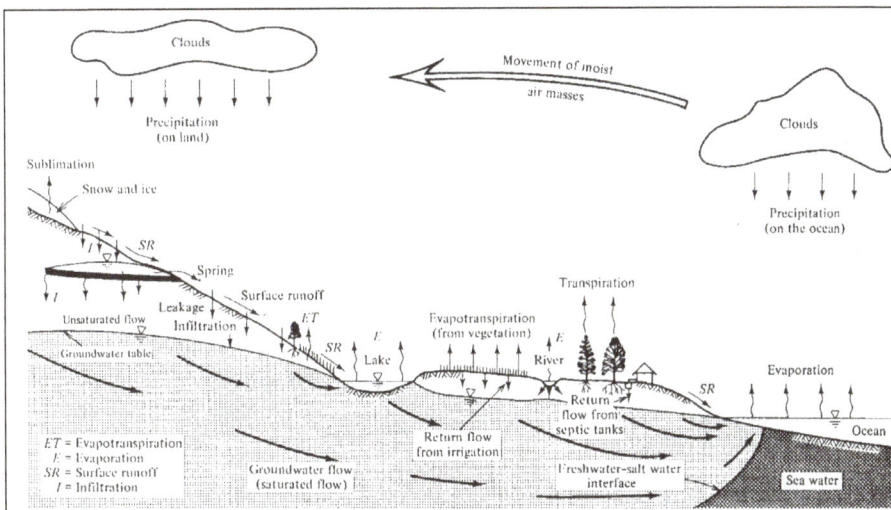

The hydrological cycle in more detail

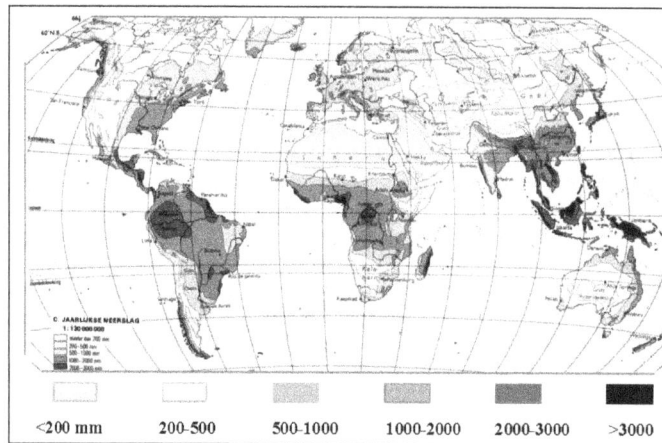

Global distribution of the yearly precipitation

Hydrological Water Balance

A basic tool of hydrology is to set up a water balance for a catchment. Figure shows the essential components and processes of a river basin. Incoming precipitation reaches the river by direct overland flow or by several slower pathways through the subsurface. Output from the catchment happens only by evapo(trans)piration and by the river discharge.

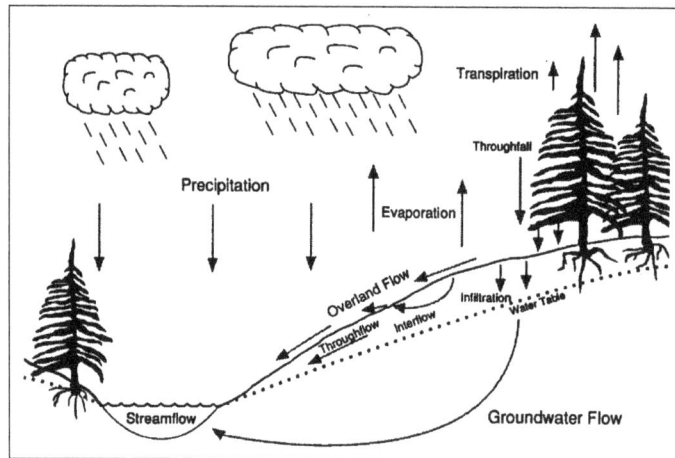

Components of the hydrological budget for a catchment

In a formal description, such a hydrological system may be seen simply as a reservoir with:

- Input P (precipitation)

- Output R (runoff)

- Output ET (evapotranspiration)

Mass conservation then requires that input – output equals the change in storage, which mathematically can be expressed as:

$$P - R - ET = \frac{dS}{dt}.$$

This basic hydrological budget equation is simple, but the individual terms are often difficult to quantify. This is a first example of a problem which isotopes may help to solve, e.g. by providing information on the evaporation and storage terms.

A related and widely used tool of hydrology is the hydrograph analysis. This is an input/output approach, in which the hydrologic system is treated as a black box. One observes the variation of the water input, e.g. due to a rainfall event, and the response of the system in the form of a time varying river discharge. From these observations, information on the hydrological system is derived, in particular about the distribution of transit times of water through the system.

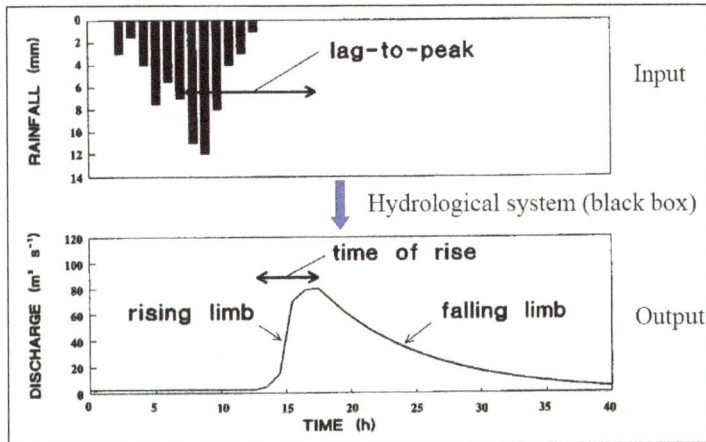

Hydrograph analysis of a hydrological system, e.g. a river catchment. Information on the transit time distribution is contained in the time lag between the peaks of the input and output as well as in the shapes of the two curves.

Several environmental tracers are very useful tools to study water residence times and transit time distributions. They are thus very well suited to complement the classical hydrograph approach. However, not only time-sensitive environ- mental tracers are useful in this context. Tracers such as stable isotopes may allow the distinction of waters of different origin (e.g. rainwater, river water, groundwater) and therefore the identification of the different pathways that the water takes from the input to the output of the system.

Properties of Water

Although water is a very common substance, it has surprisingly special properties. In most of its physical properties, water is either unique or at least close to an extreme value of the range of values occurring in natural substances. The reasons for these special properties are the polar structure of the water molecule and its ability to form intermolecular bonds (hydrogen bridges). The quite extreme physical properties of water in turn are important for the pivotal role of water in the environment.

Water is a central part of the global climate system. The ocean stores huge quantities of heat energy and transports it over large distances. Water is very efficient in storing heat due to its high heat capacity. Because the latent heats of fusion and evaporation are also very high, there is also a lot of energy transferred in the processes of melting/freezing and evaporation/condensation. Water vapour absorbs infrared radiation and thus yields the largest contribution to the natural greenhouse effect. On the other hand, clouds reflect solar radiation and thereby also have a cooling effect.

Due to its high dielectric constant and dissolving power, water is a very good solvent and plays a central role in weathering and erosion as well as the transport of the resulting material by rivers. The ability of water to dissolve and transport ions and nutrients is also of crucial importance for life. Water is the most frequent substance in the biosphere, as it forms the medium of the chemical processes of life.

Table: Physical properties of water.

Property	Comparison	Importance and consequences
Specific heat (per unit mass)	Highest of all solids and liquids except liquid NH_3	Heat transport by water movement, heat buffering
Latent heat of fusion	Highest except NH_3.	Thermostatic effect at freezing point
Latent heat of evaporation	Highest of all substances	Heat and water transfer in the atmosphere
ρ max at T > Tfreezing (~4 °C at 0%, 1 atm)	anomalous	Density stratification of lakes, facilitates freezing
ρ solid < ρ liquid	anomalous	Ice floats on water, freezing only at surface, weathering
Surface tension	Highest of all liquids	Drop formation, capillary forces, soil water retention, cell physiology
Dissolving power	Very high	Transport of dissolved substances
Dielectric constant	Highest of all liquids except H_2O_2 and HCN	High dissociation of dissolved salts
Transparency	Relatively high	Enables photosynthesis also at considerable water depth.

The peculiar temperature dependence of the density of water is of particular importance for the physics of freshwater systems, in particular lakes. It leads to a vertical density stratification of lakes and forms the basis of physical limnology. We therefore look a bit closer at water density here.

The density r of water depends on temperature T and salinity (salt concentration) S. The salinity dependence is quite complicated in detail, because the influence of each individual salt species is different. In practice, however, the salinity can be parameterised by the easily measurable electrical conductivity (although this has in principle to be done for every lake separately), and a linear relationship between salinity and density is often a useful approximation for lakes. For the ocean, very precise equations linking conductivity, salinity, and density have been developed. For a more detailed treatment of the salt dependence of water density the reader is referred to the script of the first part of this lecture. In the following, we focus on the density of pure water.

The highest density is attained at a temperature of 3.98 °C (ρ = 999.972 kg m⁻³). The function $\rho\,(T)$ for pure water at a pressure of 1013 mbar (1 atm) can be approximated by a simple quadratic equation:

$$\rho\,(T) = 999.972 - 7 \cdot 10^{-3}(T-4)^2 \text{ with T in °C and ρ in kg m⁻³}$$

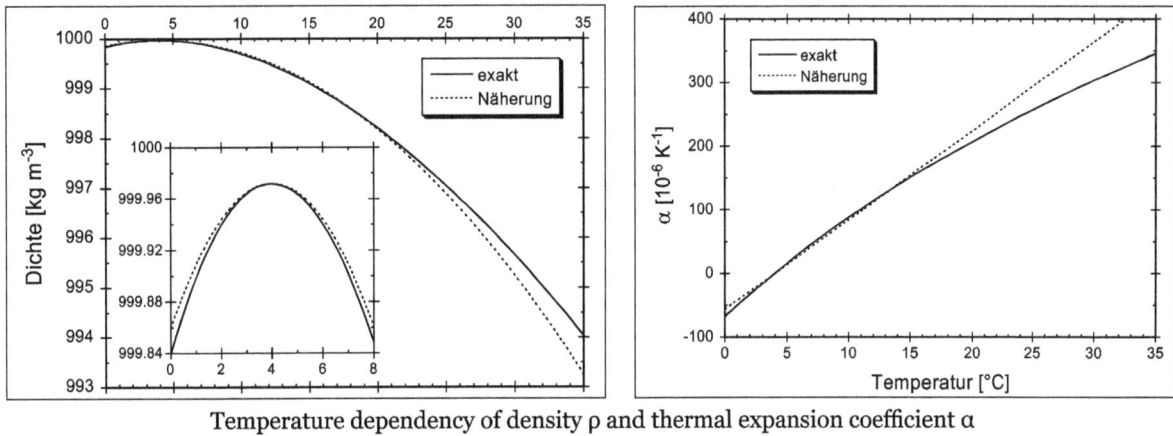

Temperature dependency of density ρ and thermal expansion coefficient α

If only density differences are important, the maximum density can be replaced simply by 1000 kg m^{-3}. The temperature dependence of the density is often described by the coefficient of thermal expansion α, which is defined as follows:

$$\alpha = -\frac{1}{\rho}\left(\frac{\partial \rho}{\partial T}\right)_p$$

Thus, α describes the relative density change ($\Delta\rho / \rho$) of water per change of temperature (ΔT). A negative sign occurs because α is meant to describe the relative expansion, which is inversely related to density. It would disappear if instead of density its inverse, the specific volume v = V/m was used. Since the density decreases (v increases) with temperature above $T_{\rho max}$ = 3.98°C, α is positive in this temperature range, and negative for T < $T_{\rho max}$. At the temperature of maximum density, α vanishes.

Using the simple parabolic approximation for ρ, α becomes a linear function of T. These approximations are compared to the exact values in figure.

Surface Water: Stratification, Turbulence and Transport

Basics of Physical Limnology

Limnology is the science of fresh surface waters, in particular lakes. The word is derived from the Greek "limne", meaning lake. Limnology is mainly a biological science, investigating the structure and function of lakes and river ecosystems. However, knowledge of the physical and chemical processes is a prerequisite for an understanding of the biological processes in such water bodies.

Physical limnology addresses questions related to the physical structure and the physical processes in lakes, and is one of the major areas of the physics of aquatic systems. The classic item of physical limnology is the description of the density stratification in lakes and its seasonal dynamics. Due to the peculiar temperature dependence of the water density, water of a temperature around 4 °C is densest and sinks to the ground. As a result, a deep-water body of more or less homogeneous temperature and nearly maximal density is formed, the so-called hypolimnion. In summer, the warmer and less dense water floats at the surface, creating a stable density stratification. The upper layer is well-mixed due to the energy input from the wind. It is called the epilimnion. The layer with the strong temperature gradient between the warm surface and cold deep water is called the thermocline.

The strong density gradient in the thermocline prevents vertical mixing in summer in suffi- ciently deep lakes. Only in fall or winter, when the surface water cools to temperatures close to 4 °C, the density contrast disappears and a complete vertical convective mixing (turnover) can occur. In winter, an inverse stratification with water near the freezing point floating on the 4 °C deep water can occur. Lakes that experience a full turnover every winter are called holomictic (if two turnovers in fall and spring occur, also dimictic). Lakes which never mix completely are called meromictic.

In the ocean, there is no density anomaly as in fresh water, because the temperature of maximum density decreases with increasing salinity until it becomes lower than the also (but weaker) decreasing melting point. As a result, water with salinities higher than about 25 ‰ (ocean water has ~ 35 ‰) behaves as a "normal" fluid with respect to thermal expansion.

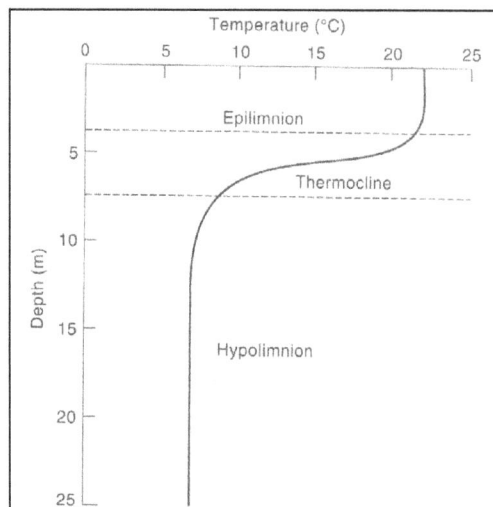

Density stratification of lakes and the resulting water layers

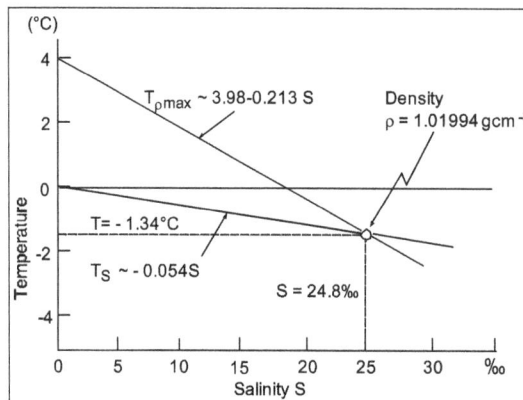

Vanishing of the density anomaly of water with increasing salinity

Fluid Dynamics: Navier-Stokes-Equation

The basis of any description of geophysical fluid dynamics, and thus the physics of water flow in aquatic systems, is the Navier-Stokes-equation. It is the equation of motion for a viscous fluid on the rotating Earth:

$$\frac{dv}{dt} = \frac{\partial v}{\partial t} + (v \cdot \nabla)v = -g\hat{z} - \frac{1}{\rho}\nabla p - 2(\Omega \times v) + \nu\Delta v$$

On the left hand side of this equation we have the acceleration of a water parcel, where the total derivative splits in two terms, describing the acceleration at fixed points in space (temporal change of the velocity field) and the change of velocity due to the spatial structure of the velocity field. The latter term is non-linear and complicates the solution this equation considerably. The right hand terms of the Navier-Stokes-equation describe accelerations due to different forces, namely: Gravity, pressure gradient force, Coriolis force, and internal friction.

Although fundamental, the Navier-Stokes-equation will not be of great immediate importance for isotope and tracer applications in aquatic systems. Much more directly accessible quantities are parameters that describe transport of tracers and other dissolved substances. We thus need to look at transport processes, which in surface waters requires a description of turbulence.

Turbulence in Surface Waters

Flow in surface waters is nearly always turbulent. This can be seen from the Reynolds number Re, which is a criterion for the occurrence of turbulence. Re is a measure of the relative importance of the non-linear and friction terms in the Navier-Stokes-equation. The non-linear term can be thought of as responsible for turbulence production, whereas friction due to molecular viscosity is the final sink of kinetic energy and thus ultimately destroys turbulence. We calculate Re by means of a scale analysis, using a characteristic velocity U and a length L to estimate the terms:

$$Re \equiv \frac{\text{Non-linear term}}{\text{Molecular friction term}} = \frac{u\dfrac{\partial u}{\partial x}}{\nu\dfrac{\partial^2 u}{\partial x^2}} \sim \frac{U\dfrac{U}{L}}{\nu\dfrac{U}{L^2}} = \frac{UL}{\nu}$$

Experiments of Reynolds showed that the transition from laminar to turbulent flow occurs for Re > Re$_c\sim$2000. For the open ocean we can chose $U = 0.1$ m s^{-1} and $L = 10^6$ m as typical scales. With $\nu \sim 10^{-6}$ m² s^{-1} we obtain Re $\sim 10^{11} \gg$ Re$_c$. Thus the nonlinear terms are very much larger than the internal friction term leading to turbulent flow. In lakes, the scales are somewhat smaller, but the condition for turbulent flow is still met by a wide margin.

The occurrence of turbulence means that the flow field is variable at all scales. This complicates the description of transport enormously, because in principle one would need to know the flow field exactly even at the smallest scales to follow the trajectory a dissolved particle. However, this situation can be remedied by the following treatment:

- Split the flow field in a mean current and statistical fluctuations.

- Treat the transport by the mean current as advection (transport by ordered flow).

- Treat the effect of small scale turbulence as diffusion (transport by unordered flow) As a result, a new transport parameter is introduced, the so-called turbulent (eddy) diffusivity K. Since turbulence often is not isotropic, i.e. it depends on the direction, we actually need a set of turbulent diffusion coefficients, which can be summarised by a tensor K.

Transport in Surface Waters

The equations describing transport processes in a turbulent medium are summarised in the following. First, we look at flux densities, i. e. the mass of tracer transported through a unit cross-sectional area per unit time (with c = concentration):

Advective flux:

$$F_{ad} = cv$$

Molecular-diffusive flux (Fick's 1st law):

$$F_{diff} = -D_m \nabla c$$

Turbulent-diffusive flux:

$$F_{turb} = -K \cdot \nabla c$$

Because turbulent diffusion (the components of K) is usually much larger than molecular diffusivity, the molecular diffusion can be neglected, and the total transport can be written as:

$$F_{tot} = cv - K \cdot \nabla c$$

To obtain the concentration change at a given location, the balance of the fluxes into and out of a control volume has to be calculated. The mass balance of tracer in the control volume is given by (M: tracer mass, J: source of tracer mass per volume and time):

$$\frac{\partial M}{\partial t} = \int_V \frac{\partial c}{\partial t} dV = -\int_A F_{tot} \cdot dA + \int_V J dV$$

The minus sign for the surface integral is because dA points outwards. According to the law of Gauss the surface integral of F equals the volume integral of the divergence of F:

$$\int_A F_{tot} \cdot dA = \int_V \nabla \cdot F_{tot} dV$$

Using equation $\int_A F_{tot} \cdot dA = \int_V \nabla \cdot F_{tot} dV$, it follows from equation:

$$\frac{\partial M}{\partial t} = \int_V \frac{\partial c}{\partial t} dV = -\int_A F_{tot} \cdot dA + \int_V J dV$$

$$\frac{\partial c}{\partial t} = -\nabla \cdot F_{tot} + J = -\nabla \cdot (cv - K \cdot \nabla c) + J$$

When the same mass balance argument is applied to the water mass itself (using water density ρ instead of the tracer concentration c), the important continuity equation can be derived for an incompressible fluid without sources or sinks:

$$\text{div } v = \nabla \cdot v = 0$$

This can be used to slightly simplify the transport equation $\dfrac{\partial c}{\partial t} = -\nabla \cdot F_{tot} + J = -\nabla \cdot (cv - K \cdot \nabla c) + J$ to its final form:

$$\frac{\partial c}{\partial t} = -v \cdot \nabla c + \nabla \cdot (K \cdot \nabla c) + J$$

Measurement of tracer concentrations and their spatial and temporal variability can be very useful to determine the unknown system parameters in this transport equation, i.e. the advective flow velocity v, the turbulent diffusivities (the components of K), and possible sources or sinks J. The tensor K can often be reduced to two components, namely the vertical turbulent diffusivity K_z and the horizontal turbulent diffusivity $K_h = K_x = K_y$. This is because the main anisotropy in surface waters is between the vertical and horizontal directions, due to the vertical density stratification and simply the different scales.

Measurement of Turbulent Diffusion Coefficients

Conservative tracers are especially suitable to determine turbulent diffusivities. Such applications play a minor role in this course, which mainly focuses on environmental tracers.

Assuming a constant vertical diffusion coefficient K_z and purely turbulent diffusive transport for a conservative tracer, the transport equation $\dfrac{\partial c}{\partial t} = -v \cdot \nabla c + \nabla \cdot (K \cdot \nabla c) + J$ reduces to:

$$\frac{\partial c}{\partial t} = K_z \frac{\partial^2 c}{\partial z^2}$$

This equation has analytical solutions for specific initial and boundary conditions. E.g. starting with a tracer distribution in the form of a δ-peak (tracer only at one specific depth), the solution is given by a Gauss curve with a width σ that evolves with time as follows:

$$\sigma = \sqrt{2K_z t}$$

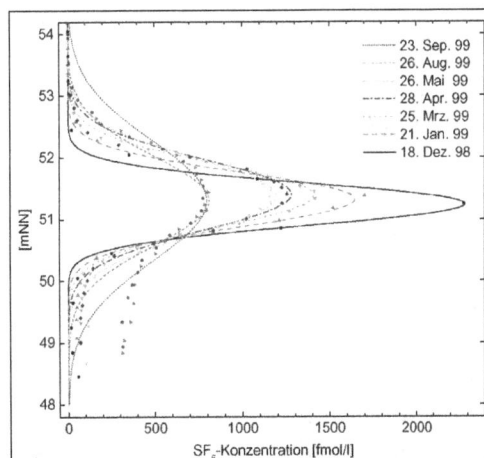

This equation can be used to determine K_z by observing the widening of a initially narrow tracer distribution over time.

Determination of the vertical turbulent diffusion by fitting Gauss curves to the observed distributions of the artificially released tracer SF_6 in the mining lake Merseburg-Ost 1b. Due to an extremely high density stratification in this lake, a very small value of $K_z \sim 10^{-8}$ m^2 s^{-1} was found, which is nearly at the level of molecular diffusion.

The same method can also be applied to horizontal turbulent mixing. The evaluation is a bit more complicated because asymmetrical 2-dimensional tracer distributions have to be interpreted in terms of widening Gauss distributions along two major axes. Furthermore, advection cannot be neglected in the horizontal direction, thus the centre of mass of the distribution moves over time. A schematic figure of the evolution of horizontal tracer distributions in a lake is shown in figure.

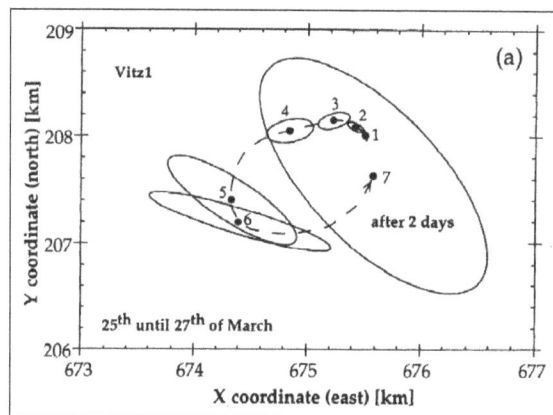

Determination of the horizontal turbulent diffusion by observing distributions of the artificially released dye Uranin Lake Lucerne (Switzerland). An idealised temporal evolution of an experiment is shown. Values of K_h between 0.02 and 0.3 m^2 s^{-1} were found.

Groundwater (Hydrogeology): Potential, Flow and Transport

Basics of Hydrogeology

By groundwater we mean the water in the subsurface that complete fills the pore space in a soil or rock matrix. The uppermost zone of the subsurface, where the pore space is only partly filled with water, and the remainder contains air, is referred to as the vadose or unsaturated zone. This is the realm of soil physics. Hydrogeology and groundwater hydrology treat mainly the saturated zone. The zones in the subsurface are shown in figure.

A geological formation that is filled with water and has a connected pore space in which the water can circulate relatively easily is called aquifer. If the formation is (nearly) impermeable for water, it is called aquitard. There are different types of porous media in which ground- water can circulate. The most important in practice are fine-grained sediments such as gravel, sand, and sandstone. The pores in such media are well-connected, and the system is relatively homogeneous. In hard rocks, such as granite, the water can only move in fractures ("fractured rock" aquifers). A third type of porous medium is karst, where water flows in relatively large openings or cavities. This occurs in limestone (calcite rock), which can be partly dissolved by the water.

An important quantity to characterise porous media is the porosity, which is defined as:

$$\theta = \frac{V_{pores}}{V_{total}}$$

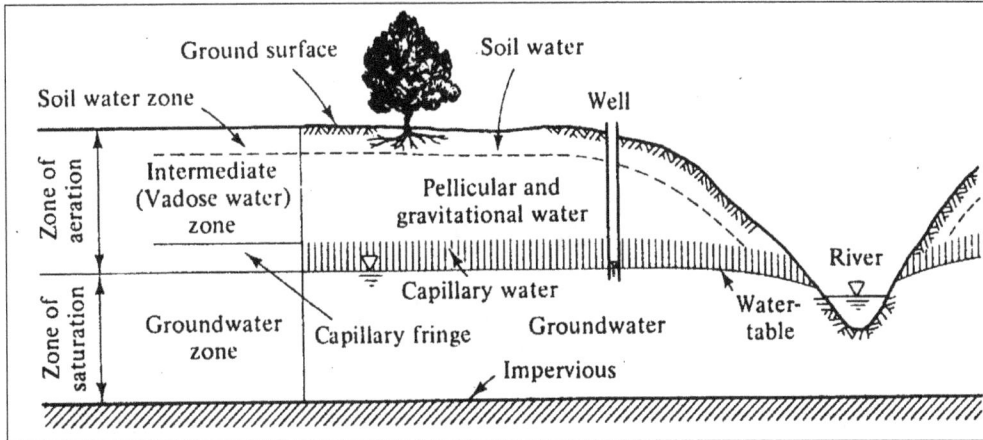

Zones in the subsurface. The soil water zone includes the root zone. The water content is subject to strong daily and seasonal fluctuations. Depending on the ratio of infiltration and evaporation, the water moves downwards or upwards. In the vadose zone, immobile water is held by capillary forces. In the case of infiltration, the water moves downwards due to gravity. The capillary fringe extends from the water table of the saturated zone upwards up to the maximum capillary rise height. This height depends on the soil type and is small in coarse material but can reach up to 3 m in very fine material with small pores. The pressure in the capillary fringe is below the atmospheric pressure, as water is sucked up by capillary forces. The groundwater zone or saturated zone is the zone beneath the groundwater table. The groundwater table is defined as the level in the saturated zone where the hydrostatic pressure equals the atmospheric pressure.

There are two different types of aquifers. Unconfined or phreatic aquifers are in direct contact with an overlying unsaturated zone. The water table, i.e. the boundary between saturated and unsaturated zone, is free to move depending on the amount of water that is present. The thickness of the water bearing zone, i.e. the aquifer itself, varies with fluctuations of the water table. The pressure at the water table equals the atmospheric pressure. If a well is drilled into an unconfined aquifer, the water level in the well equals the level of the water table in the aquifer.

In contrast, confined aquifers are overlain by an impermeable layer (aquitard). Such aquifers are always filled with water to the top, so there thickness is constant. There is no real water table, since the pressure at the upper boundary of the aquifer is larger than atmospheric. However, if a well is drilled into a confined aquifer, the water will rise to a level where the pressure equals the atmospheric pressure. The imaginary surface connecting the water levels in wells drilled into a confined aquifer is the so-called potentiometric surface and reflects the pressure distribution in the aquifer. If this surface intersects the land surface, the water has the potential to rise to the surface and flow out freely from a well. Such wells are called artesian wells.

Sedimentary basins usually contain a stack of aquifers and aquitards, with an unconfined aquifer at the top and often several confined aquifers below. Typically aquifers are formed by relatively coarse and permeable sand or gravel deposits, whereas very fine clay sediments form nearly impermeable aquitards. However, the distinction between aquifers and aquitards is not absolute, as even aquitards have a non-zero, albeit very low, hydraulic conductivity.

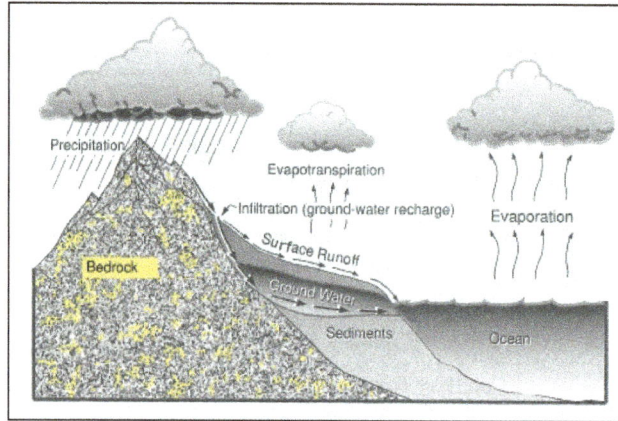

Typical layering in sedimentary basins with a sequence of aquifers and aquitards. The uppermost aquifer is usually unconfined the lower aquifers are confined by the overlying aquitards. The potentiometric surface of the lower confined aquifer is often higher than the free water table in the unconfined aquifer, implying that the (very slow) flow through the aquitard is upwards.

The (potential) height h of the water level in an aquifer, i.e. the height to which the water rises in a well, is a central quantity in hydrogeology. It obviously is related to the pressure in the aquifer. It is measured by observing the water level in boreholes, which are called piezometers, hence this height is sometimes called piezometric head. We will use the more common term "hydraulic head".

To understand the physical meaning of the hydraulic head, we consider the total energy of a water parcel in the porous medium at a certain level z above an arbitrary reference level z = 0. This energy is expressed relative to the energy of a free (not pressurised, outside the porous medium) water parcel at the reference level, and is given by:

$$E_{tot} = E_{int} + E_{pot} + E_{kin} = pV + mgz + \frac{1}{2}mv^2$$

In groundwater the flow velocity is usually very small, so that the kinetic energy term can be neglected. Dividing by the weight mg = ρVg of the water parcel, a quantity with the dimension of a length is obtained, which is the hydraulic head, i.e. the energy density of the water with respect to weight (not mass or volume):

$$h = \frac{p}{\rho g} + z$$

The hydraulic head is the height to which the water in an aquifer can rise due to its inner energy (pressure) above its current height z. It can easily be measured by allowing the water to rise in a piezometer. The situation in unconfined and confined aquifers and the terms contributing to h are illustrated infigure.

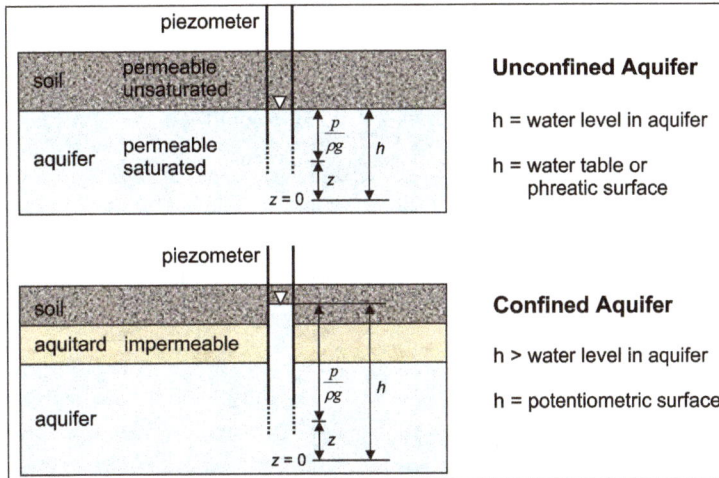

On the definition of the hydraulic head in unconfined and confined aquifers

Darcy's Law

Henry Darcy, a French engineer and scientist, studied the water flow through homogeneous sand columns in 1855 and 1856 in experiments of the type shown in figure. In Darcy's experiment, a column is completely filled with water and the hydrostatic pressure on either side is kept constant and given by the height h of the water columns in the in- and outlet tubes. Darcy found that the discharge Q, i.e. the volume of water flowing through the column per unit time is proportional to the cross section A of the sand column and to the head difference ($\Delta h = h_1 - h_2$), and inversely proportional to the length L of the column.

$$Q = KA\frac{h_1 - h_2}{L} = KA\frac{\Delta h}{L}$$

If we define the specific discharge q as the flow per cross-sectional area, q = Q/A, and the gradient of the hydraulic head along the column dh/dl = -Δh/L (negative because the head decreases in the direction of flow), Darcy's law can be written as follows:

$$q = -K\frac{dh}{dl}$$

The specific discharge q has the dimension of a velocity [m s^{-1}] and is called the Darcy velocity. The dimensionless gradient of the hydraulic head describes the pressure gradient that drives the flow trough the porous medium. The constant of proportionality K [m s^{-1}] is called the hydraulic conductivity. Its inverse parameterises the flow resistance of the medium and the internal friction of the fluid. Darcy's law is analogous to Ohm's law of electrical resistivity, with q being equivalent to the current density j, dh/dl to the voltage (potential gradient) and K to the electrical conductivity, i.e. the inverse of the specific resistance ρ.

The Darcy velocity is a hypothetical velocity, because not the entire cross-section A is really permeated. Effectively only the cross-sectional area $A_{eff} = A\,\theta$ is available for the flow, thus the mean linear velocity $v = q/\theta$ is larger than the Darcy velocity. This is the velocity that determines how long it takes for a water parcel to travel a certain distance in the aquifer.

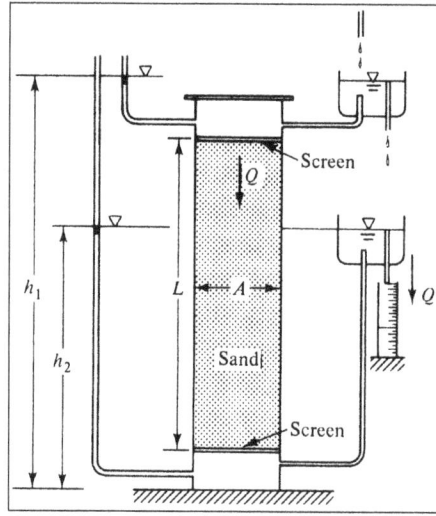

Setup of the Darcy experiment

Darcy's law can be generalised to flow in a 3-dimensional porous medium, if K is replaced by a tensor K of hydraulic conductivity. This is because the hydraulic properties of the medium are usually anisotropic, i.e. dependent on the direction. The 3-D version of Darcy's law reads:

$$q = - K \cdot \nabla h.$$

In an isotropic medium, the hydraulic conductivity is the same in all directions, the tensor K reduces to a constant, and the water flow is always parallel to the gradient of the hydraulic head. In an anisotropic medium, this is not the case. The anisotropy of porous media results from the layering of the sedimentary deposits. Usually the layers lie more or less horizontally, hence the horizontal conductivity is larger than the vertical. Choosing the x- and y-coordinates along the layers and the z-coordinate perpendicular to the layering, one can assume that K is diagonal and consists of only three components K_x, K_y, and K_z. Often the two horizontal components are further assumed to be equal and larger than the vertical, i.e. $K_x = K_y = K_h > K_z$. In this case, the horizontal flow direction is always parallel to the gradient and perpendicular to the isolines of the hydraulic head.

Hydraulic Conductivity and Permeability

The central quantity in Darcy's law is the hydraulic conductivity K. It describes the hydraulic properties of the porous medium, but also depends on the properties of the fluid (viscosity). To separate these two influences and really isolate the properties of the medium, one can use the (intrinsic) permeability, which is defined as:

$$k = \frac{K\mu}{\rho g} = \frac{Kv}{g}$$

Where μ is the dynamic, v the kinematic viscosity and ρ the density of the fluid. For water the conversion factor v/g is about 10^{-7} m s ($v \sim 10^{-6}$ m² s^{-1}, $g = 10$ m s^{-2}). The permeability k has the dimension of an area (m²) and is only a property of the porous medium. It can be parameterised as $k = Cd^2$, where d is the grain size (~pore size) and C describes a property of the aquifer matrix. This shows that the permeability depends strongly (quadratic) on the grain respectively pore size.

On the other hand it does not directly depend on porosity, because porosity is largely independent of grain size. Coarse and fine grained media (e.g. gravel and sand) can have the same porosity but strongly different permeability. Some typical values for grain size, porosity, and hydraulic conductivity of different sediments and rocks are listed in table.

Table: Grain size, porosity, and hydraulic conductivity of some typical aquifer and aquitard materials. The sediment type gravel, sand, silt and clay are defined by the indicated grain size ranges.

Sediment, rock	Grain size d [mm]	Porosity θ [%]	Hydraulic conductivity K [m s^{-1}]	Permeability (qualitative)
Gravel (kies)	> 2	25 - 40	$10^{-2} - 10^{2}$	Permeable
Sand	$0.05 - 2$	25 - 50	$10^{-5} - 10^{-2}$	Permeable
Silt (schluff)	$0.002 - 0.05$	35 - 50	$10^{-9} - 10^{-5}$	Half-permeable
Clay (ton)	< 0.002	40 - 70	$10^{-12} - 10^{-9}$	Impermeable
Sandstone	-	5 - 30	$10^{-10} - 10^{-5}$	Half-permeable
Crystalline rock	-	0 - 10	$10^{-13} - 10^{-11}$	Impermeable

The hydraulic conductivity can be determined by lab experiments on sediment samples obtained in drill cores, or, more usually, by pumping tests on wells. Both ways yield only local information, which is problematic in view of the high variability of the hydraulic conductivity, as shown by the values of table.

The spatial inhomogeneity of the hydraulic conductivity and other properties of the porous medium is indeed a central problem of hydrogeology and any attempts to model groundwater flow. Since natural porous media are mixtures of materials with different grain size, the permeability or hydraulic conductivity varies strongly from place to place. It is an important empirical finding that the hydraulic conductivity K is often approximately log-normally distributed. Thus, not K itself, but log(K) follows a normal distribution. This means that K varies often over several orders of magnitude, which poses the greatest difficulty for the modeling of groundwater flow.

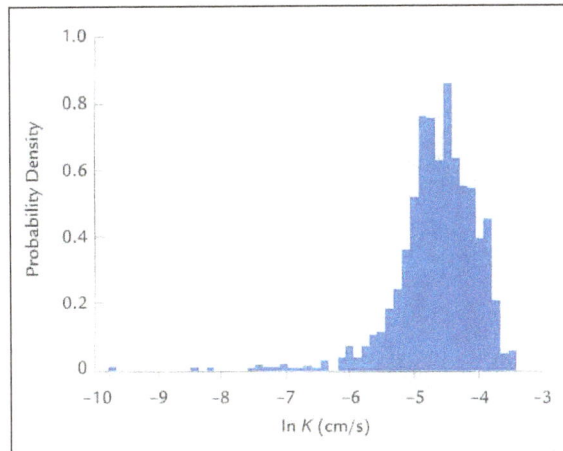

Frequency distribution of the hydraulic conductivity K measured at different points in an aquifer.
The distribution is approximately log-normal (note the log- scale of the K-axis)

Transport in Groundwater

The equations describing transport processes in a porous medium are summarised in the following. There are two main differences in comparison to the case of surface water. First, the transport happens only in the pore space, thus the factor porosity enters the equations. Second, the flow is not turbulent, so there is no turbulent diffusion.

First, we look at flux densities, i. e. the mass of tracer transported through a unit cross- sectional area per unit time (with c = concentration):

Advective flux:

$$F_{ad} = \theta c v$$

Molecular-diffusive flux (Fick's 1st law):

$$F_{diff} = -\theta D_m \nabla c$$

Dispersive flux:

$$F_{dis} = -\theta D \cdot \nabla c$$

Because dispersion (the components of the tensor D) is usually much larger than molecular diffusivity, the molecular diffusion can be neglected, and the total transport can be written as:

$$\frac{1}{\theta} F_{tot} = cv - D \cdot \nabla c$$

The mass balance for the water in a control volume implies that the divergence of the water flux q equals the source of water J_w. In contrast to surface water, where one usually has no sources or sinks of water, and this argument ends up in the well-known continuity equation stating div v = 0, in groundwater it is usually important to take sinks and sources (water withdrawal or recharge via wells) into account). Combining the continuity equation with Darcy's law yields the flow equation:

$$div\ q = \nabla\left(-K \cdot \nabla h\right) = J_w$$

Applying the same mass balance argument to the fluxes of a tracer yields the concentration change in a given control volume (or rather the fraction θ of it), which has to be equal to the divergence of the flux plus a source term J_c for the tracer concentration:

$$\frac{\partial\left(c\theta\right)}{\partial t} = -\nabla \cdot F_{tot} + J_c$$

The tracer source term can be split in two parts, one describing the local sinks and sources, the other one being due to the tracer concentration in a possible water input J_w. Assuming that the porosity is constant and inserting the fluxes:

$$F_{ad} = \theta c v$$

In equation:

$$\frac{\partial\left(c\theta\right)}{\partial t} = -\nabla \cdot F_{tot} + J_c$$

the transport equation in its final form is obtained:

$$\frac{\partial c}{\partial t} = -v \cdot \nabla c + \nabla \cdot \left(D \cdot \nabla c\right) + \frac{J_c}{\theta} + \frac{J_w}{\theta}\left(c_{in} - c\right)$$

Again, as in the case of surface water, measurement of tracer concentrations and their spatial and temporal variability can be very useful to determine the unknown system parameters in this transport equation, i.e. the advective flow velocity v, the dispersivities (the components of D), and possible sources or sinks.

Dispersion

Dispersion is an undirected contribution to the tracer transport in groundwater, which leads to a widening and smoothing of the concentration distribution. In contrast to surface waters, flow in groundwater is usually laminar, not turbulent. Thus dispersion does not result from the stochastic nature of turbulence, but from the inhomogeneity of the porous medium. Although the flow is laminar, neighbouring water parcels will experience different flow velocities and take different flow paths, leading to a distribution of flow times between two points in the medium.

Reasons for different flow times are: 1. the velocity distribution within a pore, with maximum velocity in the middle and zero velocity at the boundaries, 2. differences in the flow velocities between small and large pores, 3. differences in the length of the flow paths, forced by the geometry of the medium, 4. zones of different hydraulic conductivity, 5. the large-scale inhomogeneous layering of the medium. Points 1 to 3 are referred to as mechanical dispersion, due to the microscale properties of the porous medium. Points 4 and 5 are referred to as macrodispersion, due to inhomogeneity at larger scales. These effects are illustrated in figure.

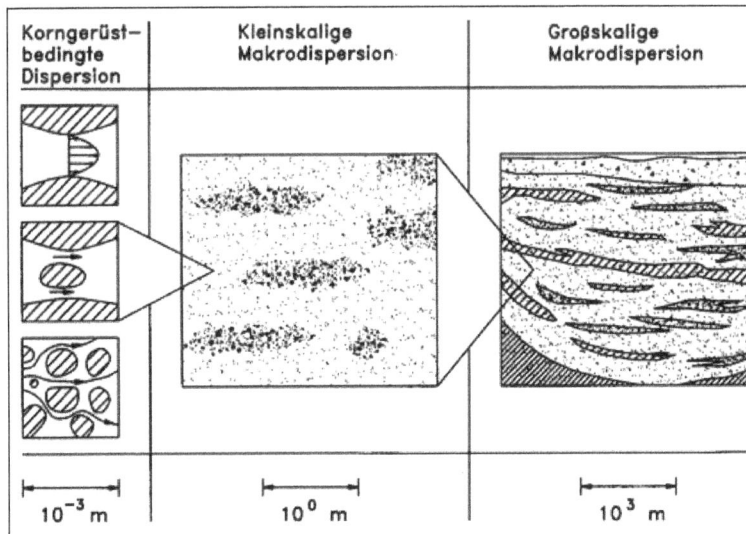

Effects contributing to dispersion on different scales

Dispersion has the same effect on tracer distributions as molecular and turbulent diffusion, and it is also mathematically described analogously. In contrast to the molecular diffusivity, but analogous to the turbulent diffusivity, dispersion is anisotropic and has to be described by a tensor D. A special property of dispersion is that its strength is not only dependent on the geometric direction in space, but also on the direction of the flow. Dispersion is usually stronger in the direction of flow (longitudinal) than perpendicular to it (transverse).

The tracer flux by dispersion is described by equation $F_{dis} = -\theta D \cdot \nabla c$ in analogy to diffusion. The dispersion tensor D can often be reduced to two components, namely the longitudinal dispersion coefficient D_L in the direction of flow and the transversal dispersion coefficient D_T in the perpendicular direction. These dispersion coefficients do not depend on the particular substance that is transported, but on the flow and on the properties of the porous medium.

These two influences can be separated by the following approximation:

$$D_L = \alpha_L u \quad and \quad D_T = \alpha_T u$$

where u is the linear velocity and α_L and α_T are the longitudinal and transverse dispersivities, respectively. The two dispersivities have the dimension of a length and are purely properties of the aquifer matrix. Typically the transverse dispersivity is an order of magnitude smaller than the longitudinal dispersivity.

References

- What-is-the-water-hydrologic-cycle: worldatlas.com, Retrieved 31 March, 2019

- Hydrology, water-cycle: nwrfc.noaa.gov, Retrieved 14 July, 2019

- Hydrological-Cycle: bhattercollege.ac.in, Retrieved 17 May, 2019

- Kuchment, doc, water, functions, library, ecoservices, Programs: biodiversity.ru, Retrieved 19 April, 2019

- Water-balance-meaning-components-and-types-hydrology-geology, water-balance, hydrology: geographynotes.com, Retrieved 5 February, 2019

- Water-balance-estimation, further-resources-water-sources-software, module-4-sustainable-water-supply, sswm-university-course: sswm.info, Retrieved 26 July, 2019

Chapter 5

Drainage and Watersheds

The removal of sub-surface or water from an area through artificial or natural means is termed as drainage. A watershed, also known as a drainage basin, refers to a piece of land where rain or snow collects and flows off into a common outlet. This chapter has been carefully written to provide an easy understanding of the varied facets of drainage, drainage systems, watershed and watershed management.

Drainage

Drainage is the removal of excess surface and subsurface water from the land to enhance crop growth, including the removal of dissolved salts from the soil. Drainage is necessary for successful irrigated agriculture because it controls ponding, waterlogging and salinity. Drainage can be either natural or artificial. Most areas have some natural drainage; this means that excess water flows from the farmer's fields to swamps or to lakes and rivers. However, the natural drainage is often inadequate to remove excess rainfall during extreme rainfall conditions or to remove the extra water or salts brought in by irrigation. Under these conditions, an artificial or man-made drainage system is required.

Need for Drainage

Drainage for Agriculture

Objectives of drainage - The four main objectives of drainage in agricultural land are:

- Drainage to prevent or reduce waterlogging.
- Drainage to control salinity, or
- Drainage to make new land available for agriculture.
- Drainage to sustain the land and water resources.

Water Balance

Agriculture depends on the availability of water. In humid regions, the main source of water is rainfall, in arid or semi-arid regions supplemented by irrigation. To apply irrigation water to a crop, water has to be diverted from a river or lake or from the groundwater reservoir. The amount of water diverted has to be greater that the quantity required by the crops because the diverted water will leave the area not only as evapotranspiration by the irrigated crop, but also as evaporation, seepage and operational spills from the irrigation canal system, as tail water runoff from irrigated fields, and as deep percolation. In the field, irrigation water together with any rainfall, will be partly stored on the soil surface and partly infiltrate in the soil.

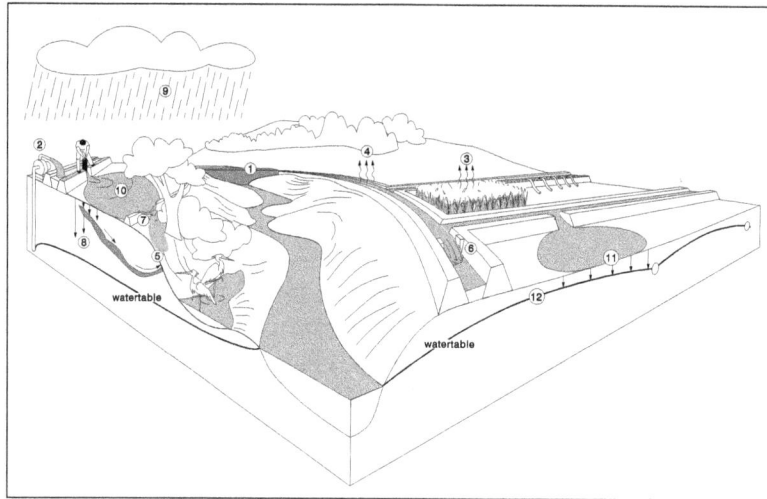

The water balance in an irrigated area

Water Ponding

When rain or irrigation continues, pools may form on the soil surface, and this excess water needs to be removed. This standing water on the soil surface is called ponding water.

Water Logging

Part of the water that infiltrates the soil will be stored in the pores and used by the crop and part of the water will be lost as deep percolation. When the percolating water reaches that part of the soil which is saturated with water, the watertable will rise. If the watertable reaches the root zone, the plants may suffer. The soil has become waterlogged.

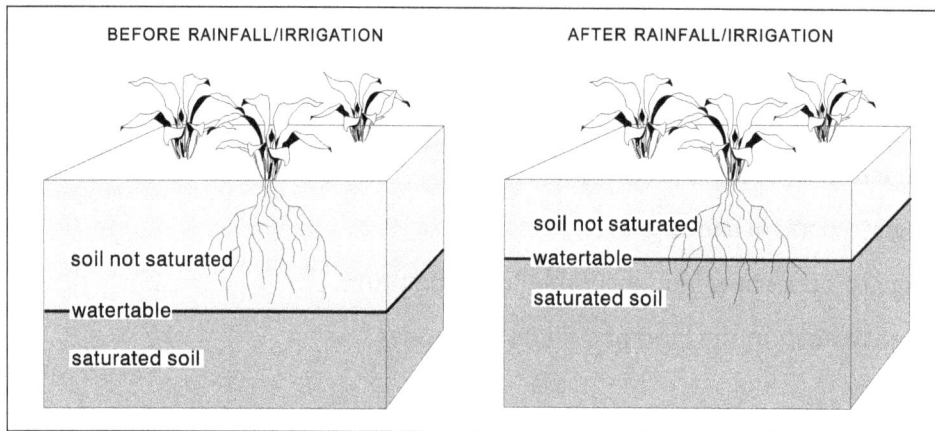

After rainfall or irrigation the watertable may rise and reach the root zone

Salinisation

Drainage is needed to remove the excess water and to control the rise of the watertable. Even in irrigation water of very good quality there are salts, thus bringing irrigation water to a field means also bringing salts to the same field. The irrigation water is used by the crop or evaporates directly from the soil. The salts, however, are left behind. This process is called salinisation.

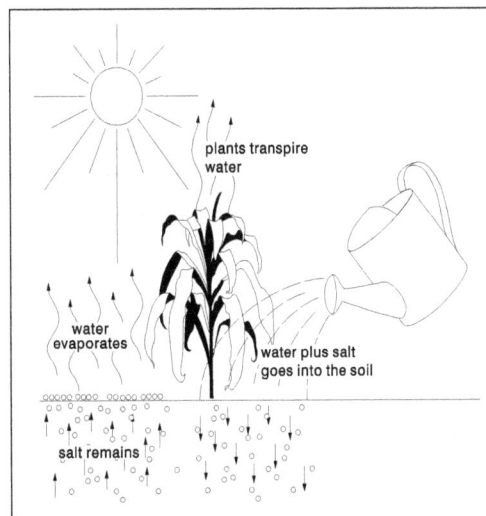

Irrigation water brings alts into the soil

If these salts accumulate in the soil, they will hamper crop production. Some crops are more tolerant to salts than others. The highly tolerant crops can withstand a salt concentration in the root zone up to 10 dS/m, the moderately tolerant crops up to 5 dS/m and the sensitive crops up to 2.5 dS/m. To grow more sensitive crops, drainage is needed to remove these salts.

Thus drainage is used to control ponding at the surface, to control waterlogging in the soil and to avoid salinisation, and may be defined as:

Drainage is necessary for successful irrigated agriculture because it controls ponding, waterlogging and salinity. Drainage can be either natural or artificial. Most areas have some natural drainage; this means that excess water flows from the farmer's fields to swamps or to lakes and rivers. However, the natural drainage is often inadequate to remove excess rainfall during extreme rainfall conditions or to remove the extra water or salts brought in by irrigation. Under these conditions, an artificial or man-made drainage system is required.

Drainage to Control Water Ponding

Surface Drainage

To remove excess (ponding) water from the surface of the land we use surface drainage. This is normally accomplished by shallow open field drains. In order to facilitate the flow of excess water towards these open drains, the field is usually given an artificial slope by means of land shaping or grading.

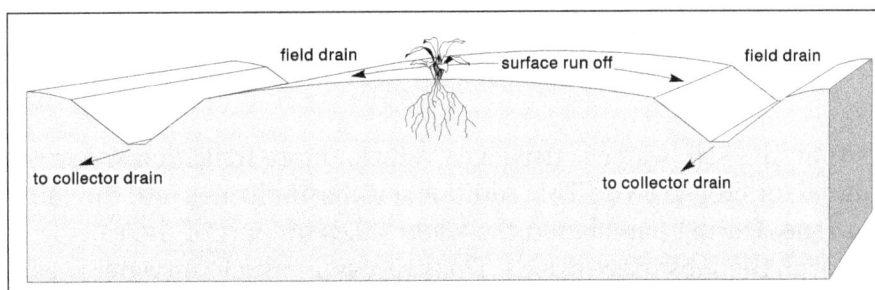

Surface drainage to remove excess water from the land surface

Drainage to Control Waterlogging

Subsurface Drainage

To remove excess water from the root zone we use subsurface drainage. By subsurface drainage we control the watertable, and excess water is removed from the underground by gravity through open or pipe drains installed at depths varying from 1 to 3 m.

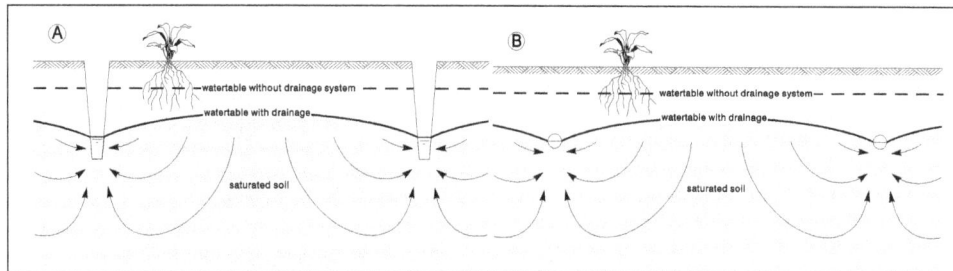

Field drains for subsurface drainage may be open (A) or pipe (B)

Tubewell Drainage

Tubewell drainage is a special type of subsurface drainage where excess water is removed by pumping from a series of wells drilled into the ground to a depth of several tens of metres. The pumped water is then discharged into open surface drains.

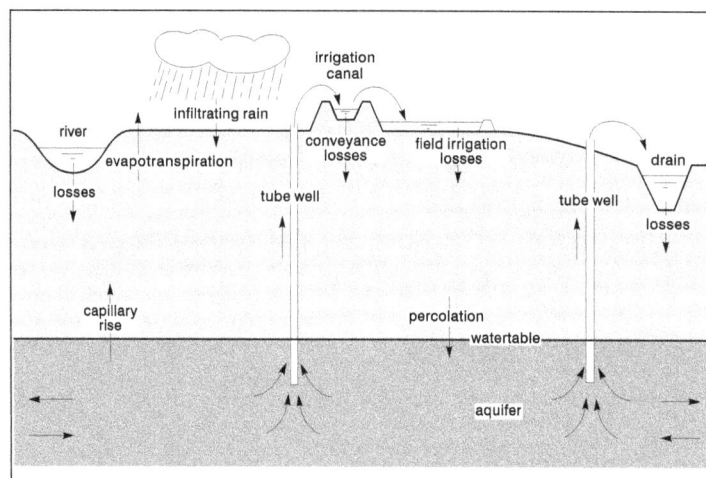

Tubewell drainage is a technique of controlling the watertable by removing the excess water from the (more permeable) underground

Drainage to Control Salinization

Leaching

To remove salts from the soil, water is used as a vehicle: more irrigation water is applied to the field than is required for crop growth. This additional water infiltrates into the soil and percolates through the root zone. During percolation the water takes up part of the salts from the soil and removes these through the subsurface drains. This process, in which the water washes the salts out of the root zone, is called leaching.

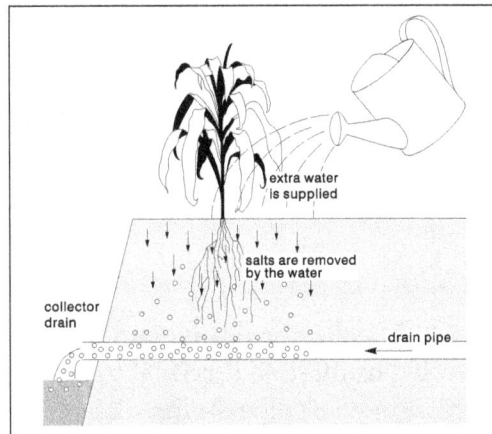
Extra irrigation water is applied to remove salts from the rootzone

The additional water required for leaching must be removed from the root zone by means of drainage, otherwise the water table will rise and this will bring the salts back into the root zone. Thus salinity control is achieved by a combination of irrigation and drainage measures.

Drainage Systems

A drainage system is an artificial system of land forming, surface and subsurface drains, related structures, and pumps (if any), by which excess water is removed from an area.

Components of a Drainage System

A drainage system can be divided into three components:

- Field drainage system

- Main drainage system

- Outlet

A field drainage system is used to avoid ponding water and to control the water table in the field. The main drainage system is used to convey the water away from the farm area. And the outlet is the point of safe disposal of the drainage water.

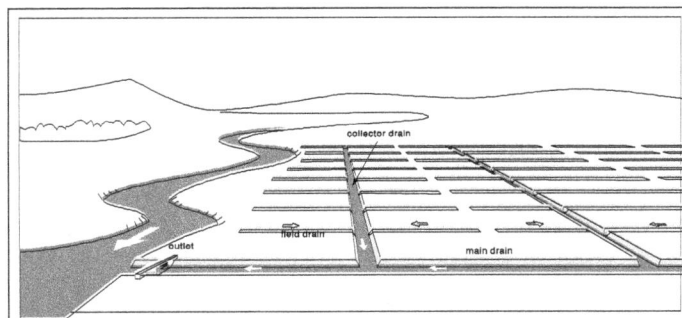
Schematic lay-out of a drainage system

Field Drainage System

For farmers, the field drainage system is the most important part of a drainage system. It controls waterponding and waterlogging in his field. It can be a surface drainage system (to remove ponding water from the surface of the land), a subsurface drainage system (to control the water table in the soil) or a combination of these two.

Main Drainage System

The main drainage system consists of collector drains and a main drain. A collector drain is a drain that collects water from the field drains and carries it to the main drain for disposal. As field drains, collector drains may be either open or pipe drains. The main drain is the principal drain of an area, it receives water from collector drains, diversion drains, or interceptor drains (= drains intercepting surface or groundwater flow from outside the area), and conveys this water to an outlet for disposal outside the area. The main drain is often a canalized stream which runs through the lowest parts of the agricultural area.

Outlet

When agricultural lands are located along rivers, lakes, estuaries, or coastal areas, dikes can protect them from being flooded. To enable the drainage of excess water from the protected area, the dikes are provided with outlet structures. These can be sluices with doors, gated culverts, siphons, and pumping stations. The water levels of the canals, rivers, lakes, or seas that receive this water may vary, because of tides, for instance. When the outer water levels are high, drainage might be temporarily restricted. This means that the drainage water accumulating inside the protected area has to be stored - in the soil, in ditches, in canals, and in ponding areas.

Gravity Outlet or Pumping Station

The outlet can be a gravity outlet structure or a pumping station. A gravity outlet structure is a drainage structure in an area with variable outer water levels, where drainage can take place by gravity when outside water levels are low. In delta areas, drainage by gravity is often restricted to a few hours per day during low tide. In the upstream regions of a river, drainage by gravity can be restricted for several weeks, during periods of high river discharges. A pumping station is needed in areas where the required water levels in the drainage system are lower than the water level of the river, lake or sea.

Outlet

The location of the outlet, where drainage water is discharged into a river, lake, or sea, influences the layout and functioning of the drainage system. To ensure the uninterrupted discharge of water throughout the drainage season, the outlet should not be blocked by a sand bank or vegetated flats, nor should it be at the inner curve of a river, where sedimentation occurs. At the outlet, the main drainage canal usually cuts through the natural river embankment or the dike. To prevent flooding of the agricultural area, the outlet is usually fitted with a sluice, which can be closed when the outside water level is too high. The sluice should be near the lowest part of the area to be drained. Soil conditions at such a location, however, may cause foundation problems, and the sluice may have to be moved to a higher, more suitable, location.

Location

To avoid damage if there is a change in the course of the river or coast line, sluices are built at a certain distance from the river or sea. The entire length of the main canal reach downstream of the sluice and part of the river embankment or coast must be protected against erosion.

Access

To operate and maintain the gates properly, it is essential that the sluice is accessible throughout the year. The cost of constructing and maintaining an all-weather access road may influence the choice of a site for the drainage outlet.

If the hydraulic gradient over the outlet sluice is insufficient to discharge all drainage water within a selected period (3 or 5 days), a pumping station may be added to the outlet. In such a case also the cost of power supply to the pumping station influences its location.

Main Drainage

Main Drainage System

Systems of drainage canals and their related structures collect and convey excess water to prevent damage to crops and to allow farm machinery to work the land. Besides these agricultural functions, a drainage canal system may have to supply water for irrigation in the dry season, act as a means of transport for shipping, etc. A main drainage system is, in principle, a mirror-image of a main irrigation system. The main function of both systems is to convey the water. The difference is that in an irrigation system the source is the inlet point in a river, lake or reservoir from where the water is distribute to all fields in the project area and the sources of the drainage system are all these fields from where the excess water is evacuate back towards the (same) river, lake or sea. The design principles are the same: either a steady-state approach (e.g. Manning) or an unsteady-state approach (e.g. Duflow) is used to calculate the dimension of the irrigation or drainage canel. A major difference is the design capacity. For irrigation canals, the design capacity is based on the crop water requirements, a flow conditions that regularly occurs (depending on the irrigation efficiencies). For drainage canals, the design capacity is based on the excess rainfall, based on a certain frequency of occurrence, e.g. once every 1, 5 or 10 years. Thus the design flow conditions in drainage canals only sporadically occur (with a frequency of once every 1, 5 or 10 years), most of the time the system operated below the design conditions.

What are the consequences of the sporadic occurrence of the design flow conditions for the O & M of a main drainage system? Compare these O&M requirements with the requirements for the O&M of main irrigation systems.

Broadly speaking, there are two kinds of drainage canal systems:

A system to intercept, collect, and carry away water from sloping agricultural lands, both agricultural and nature areas. Most of the water in this system originates from surface runoff. It will be discharged for brief periods only, causing high flow rates and sediment transport.

Flat Agricultural Plains

A system to collect and carry away water from a relatively flat agricultural area. Here, the main source of water is precipitation on the area or irrigation. Because of surface detention and groundwater storage, water is discharged over a longer period than above. Furthermore, the flat gradient canals have little or no sediment transport capacity.

Sloping Lands

If a flat agricultural area is partly surrounded by sloping lands, the surface runoff from these lands should be intercepted and discharged to prevent inundation of the agricultural area. The extent to which drainage problems in the agricultural area are caused by this surface runoff should be determined by making a water balance of the area.

High Discharges and Erosion

Runoff from sloping lands causes two major problems in the downstream areas; (i) rainfall causes high discharges of short duration, (ii) the surface runoff causes erosion, and the related sediment transport down the steep gradient of the canals causes sedimentation in the flatter canal sections.

Both problems can be eased by a combination of the following measures:

- Planting trees and encouraging the growth of natural vegetation on steep slopes;

- Contour ploughing and terracing intermediate slopes (up to 10%). Terracing is the levelling of the slopes along the contour lines in combination with the planting of crops;

- Encouraging the growth of crops that give a soil cover during the rainy season;

- Constructing retention reservoirs in the streams to temporarily store peak runoff.

These techniques are a form of erosion control; their application greatly eases the downstream drainage problems.

In sloping areas, the main drainage system usually will be limited to the reconstruction of channel reaches and to the construction of energy dissipators.

Routing

Streams originating in sloping areas can be connected to a major river, lake, or sea along two alternative routes; (i) via an interceptor canal, which channels the water around the agricultural area to a suitable outlet, or (ii) via a canalized stream through the agricultural area.

The major advantage of the interceptor canal is that peak discharges and sediments from the sloping lands do not disturb the functioning of the drainage system in the flatter agricultural area.

Limiting Discharge Rates

It is possible to limit the required discharge capacity of a channel that transports water from sloping lands to a suitable outlet if the channel discharges from one of the following two structures:

- A retention reservoir that is filled by the peak stream flow, which is then released through a bottom outlet. As a result, the discharge peak is lower, but of longer duration;

- A regulating structure that consists of a weir of limited discharge capacity in the stream and a side weir immediately upstream of it. If the stream flow exceeds a predetermined rate, it overtops the crest of the side weir. Most of the additional stream flow then discharges over the side weir into an area where inundation or overland flow causes little damage.

Which of these two lay-outs (or an intermediate lay-out) is the best solution can usually only be decided after a reconnaissance study.

Flat Agricultural Areas

The agricultural areas that require drainage are usually coastal plains, river valleys, or plains where excess rainfall and the inefficient use of irrigation water has caused waterlogging. In coastal plains, the drainage problems are exacerbated by several hydrological features, typical of such plains, being:

Problems Related to Drainage

- Gentle hydraulic gradient of the rivers in the coastal plain, which leads to low flow velocities and the deposition of sediments;

- Tidal levels in river water levels near the sea and of saline water intrusion;

- A complicated network of river branches and ramifications, which can cause natural drains to disappear in coastal swamps giving the river or stream what is known a "bad outlet";

- Rapid changes in channel configuration that can occur after each major flood;

- Low elevation of the coastal lands with respect to the level of rivers and the sea. To prevent the inundation of the coastal plain, dikes along the rivers and the sea shore are essential.

Examples of lay-outs

To illustrate alternative lay-outs for a drainage canal system, let us consider an irrigated coastal plain that lies between sloping lands (hills) and the sea. The plain is intersected by parallel rivers and streams and by an irrigation canal system. Depending on factors such as run-off from the sloping land, construction and maintenance cost of canals, quality of drainage outlets, etc., alternative lay-outs can be considered:

- Combined system: The sloping land and coastal plain are drained by one combined system.

- Separate system for sloping lands: The sloping land and coastal plain are drained by three separate systems.

- Two drainage systems in a coastal plain: The sloping land and coastal plain are drained by four separate systems.

Combined Drainage System

Figure shows a drainage canal layout that combines the drainage system of the sloping land with that in the plain. All run-off from the sloping land is intercepted and carried away by canalized streams.

These streams, and the lateral drains along the river dikes, flow into a main drainage canal that runs parallel to the sea dike. One drainage sluice with a well-defined, stable (suitable) outlet has been planned on that drain. The other streams are dammed by the sea dike. Concentrating all the drainage water discharge through one sluice eases sedimentation problems in the outlet canal.

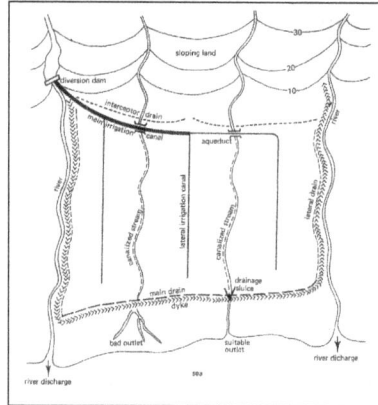

Combined system: the sloping land and coastal plain are drained by one combined system

Separate System for Sloping Land

If relatively high discharges come from the sloping lands, or if the plain is wide, intercepting and diverting streams into the nearest river is a sound alternative to the lay-out of the combined system. The streams are dammed and the interceptor drains discharge all water from the sloping lands through two sluices into the rivers. As a result, the coastal plain has a separate drainage system that discharges precipitation, unused irrigation water, and groundwater inflow. Drainage has been decentralized into three independent systems: two for the sloping land and one for the coastal agricultural area.

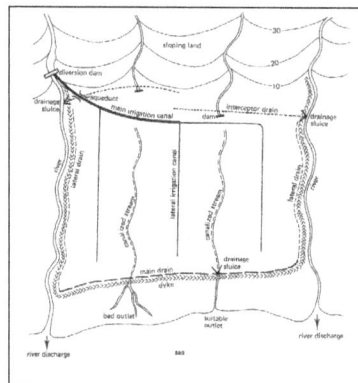

Separate system for sloping lands: the sloping land and
coastal plain are drained by three separate systems

Two Drainage Systems in a Coastal Plain

The transport of mud and sand along a coastline often blocks the outlets of all minor streams into the sea, and dredging may be needed to maintain a sufficient depth at the river mouths. Under such circumstances, none of the stream mouths is suitable as a drainage outlet. Water that is collected by the main drain along the coastal dike is then discharged into the nearest river. Figure shows four separate drainage canal sub-systems: two for the sloping lands and two for the coastal plain.

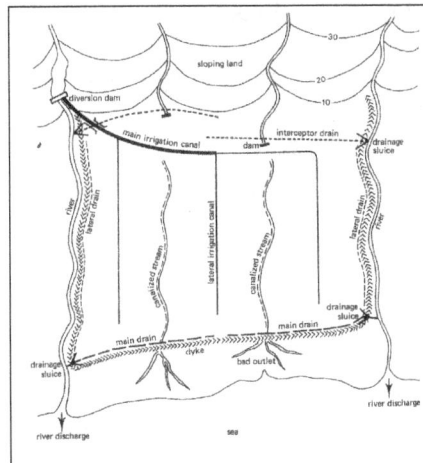

Two drainage systems in a coastal plain: the sloping land and coastal
plain are drained by four separate systems

Planning the Layout of a Main Drainage System

Location of Drainage Channels and Related Structures

To determine the location, hydraulic properties, visual characteristics, and condition of existing channels, planned canals, and related structures, one needs a 1:10 000 scale topographical map with a contour interval of 0.50 m or less, and a 1:10 000 controlled photo mosaic. Maps, especially in flat topography should be field checked. This step should be done in the earliest planning stage to avoid the need for major revisions later. The following information is needed to plan a canal system:

1. Layout of the drained area and the junctions of existing streams and all flow control points. Drainage areas should also be delineated for the "land level units";

2. Approximate profiles in existing channels, showing the elevation of the channel bottom, low bank, points of natural low ground away from but subject to drainage into the channel, and elevation and dimensions of all structures in or over the channel. The condition and serviceability of all structures should be recorded. Adequate survey data are needed for all structures to compute the discharge capacity for each;

3. Representative channel and valley cross sections for each hydraulic or economic reach. Additional cross sections should be taken as needed for a reliable estimate of: quantities of excavation and land clearance, damage evaluation in the plain or valley because of high water levels, and to permit the computation of storage in flood plains, ponds and marshes;

4. Manning's roughness coefficient 'n' for each existing channel. Even if channel elements are very uniform, the n value should be estimated for each 1-km reach;

5. Location and elevation of all soil investigation sites along the proposed canals. To determine the maximum permissible velocities and bank slopes, soil investigations should extend to a depth of at least 3 m below the anticipated future canal bottom.

6. Landscape character and use patterns along major existing and anticipated drains. Data must include: scenic views, area and density of brush and trees, and isolated but valuable trees;

7. Location and ownership of boundary lines in the vicinity of all probable canals and structures;

8. Other significant features that will be affected such as roads, pipelines, power and telephone lines, buildings, wells, cemeteries, and fences.

Field Investigations for Canal Alignment

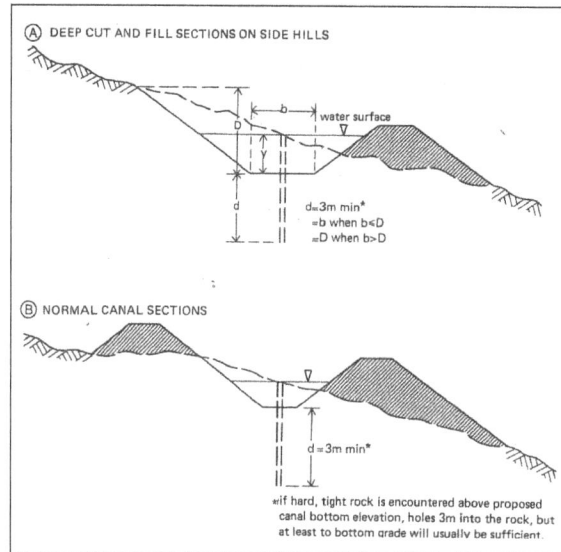

Depth of preliminary exploratory holes for canal alignment

Preparing a Preliminary Design

Based on the above information, the center line of all the canal system is drawn in pencil on the photo mosaic, showing curves, intersecting angles, and so on. Mark the stationing on these center lines with a short dash at each 100-m point.

After this preliminary design phase at the office, the canal location should be field-checked. For this check, one should walk the full length of the canal's center line, noting the following on the preliminary design drawing:

- Probable realignment of the center line;

- Points of significant breaks in the grade;

- Location of all rock outcrops or critical soil conditions;

- Approximate locations of points where more cross sections could be obtained;

- Location of significant canal junctions and places where side inlets may be needed;

- If not already visible on the aerial photo, note the location of all buildings, utilities and structures that may be affected by the drainage canal works. These include, but are not limited to, facilities that are within 100 m of the alignment and 1 m below the future canal bottom;

- Location of valuable landscapes and large individual trees adjacent to the alignment.

Following the field check, one should accurately establish the revised center line on the photo mosaic. The final alignment should be based on the previous cross sections, and geological and environmental data. Indicate on the photo mosaic where the cross sections and soil surveys were made.

Schematic Map of the Main Drainage System

Maps showing the layout of a drainage canal system must give detailed information on the location of canal reaches and related structures. Normally, this information is given on the same map that shows the irrigation canal system, roads, and the boundaries of irrigation units. To keep such maps legible, standard symbols must be used to indicate the center line of the canals and related structures. The schematic map in figure uses these symbols. It shows:

- Location of the center lines of drains and irrigation canals, numbered for each reach;

- Radii of the center lines;

- Reserve boundaries of canals and boundaries of any adjacent obstructions, roads, and land level units. The area of land level units must be shown also;

- Boundaries and number of irrigation units (if applicable);

- All structures, numbered and with position dimensioned with respect to center lines or boundaries;

- North point and scale.

Watershed

A watershed is also called a drainage basin or catchment area, is defined as an area in which all water flowing into it goes to a common outlet. People and livestock are the integral part of watershed and their activities affect the productive status of watersheds and vice versa. From the hydrological point of view, the different phases of hydrological cycle in a watershed are dependent on the various natural features and human activities. Watershed is not simply the hydrological unit but also socio-political-ecological entity which plays crucial role in determining food, social, and economical security and provides life support services to rural people.

Delineation of Watershed

Hydrologically, watershed is an area from which the runoff flows to a common point on the drainage system. Every stream, tributary, or river has an associated watershed, and small watersheds aggregate together to become larger watersheds. Water travels from headwater to the downward location and meets with similar strength of stream then it forms one order higher stream as shown in figure.

The stream order is a measure of the degree of stream branching within a watershed. Each length of stream is indicated by its order (for example, first-order, second- order, etc.). The start or headwaters of a stream, with no other streams flowing into it, is called the first-order stream.

First-order streams flow together to form a second-order stream. Second-order streams flow into a third-order stream and so on. Stream order describes the relative location of the reach in the watershed. Identifying stream order is useful to understand amount of water availability in reach and its quality; and also used as criteria to divide larger watershed into smaller unit. Moreover, criteria for selecting watershed size also depend on the objectives of the development and terrain slope. A large watershed can be managed in plain valley areas or where forest or pasture development is the main objective. In hilly areas or where intensive agriculture development is planned, the size of watershed relatively preferred is small.

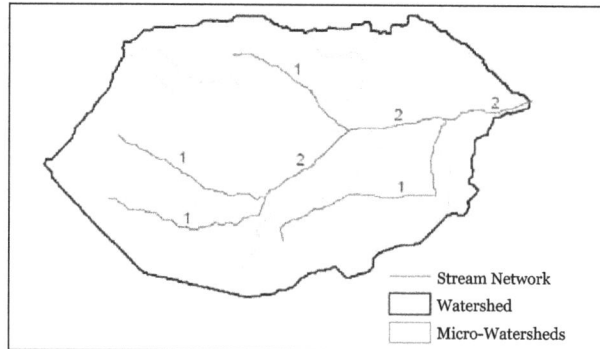

Stream network, micro-watersheds and watershed large watershed has divided into six micro-watershed based on stream order. Numbers on the stream network shows the stream order of respective stream.

Watershed Management

Watershed management implies an effective conservation of soil and water resources for sustainable production with minimum non-point resources (NFS) pollutant losses. It involves management of land surface and vegetation so as to conserve the soil and water for immediate and long term benefits to the farmers, community and society as a whole.

Catchment area is the water collecting area. "All the areas from which water flows out into a river or water pool".

Types of Watershed Management

Watershed is classified depending upon the size, drainage, shape and land use pattern:

- Macro watershed: 1000 -10,000 ha

- Micro watershed: 100 -1000 ha

- Mini watershed: 10 -100 ha

- Mille watershed: 1 -10 ha.

Objectives of Watershed Management

- Production of food, fodder, fuel.

- Pollution control.

- Over exploitation of resources should be minimized.

- Water storage, flood control, checking sedimentation.

- Wild life preservation.

- Erosion control and prevention of soil, degradation and conservation of soil and water.

- Employment generation through industrial development dairy fishery production.

- Recharging of ground water to provide regular water supply for consumption and industry as well as irrigation.

- Recreational facility.

Steps in Watershed Management

Watershed management involves determination of alternative land treatment measures for, which information about problems of land, soil, water and vegetation in the watershed is essential. In order to have a practical solution to above problem it is necessary to go through four phases for a full scale watershed management.

Phases

- Recognition phase.

- Restoration phase.

- Protection phase.

- Improvement phase.

Recognition Phase

It involves following steps:

- Recognition of the problem.

- Analysis of the cause of the problem and its effect.

- Development of alternative solutions of problem.

Restoration Phase

It includes two main steps:

- Selection of best solution to problems identified.

- Application of the solution to the problems of the land.

Protection Phase

This phase takes care of the general health of the watershed and ensures normal functioning. The protection is against all factors which may cause determined in watershed condition.

Improvement Phase

This phase deals with overall improvement in the watershed and all land is covered. Attention is paid to agriculture and forest management and production, forage production and pasture management, socio economic conditions to achieve the objectives of watershed management.

Water Resources Development Plan

Water resource management plays a vital role in sustainable development of watershed which is possible only through the implementation of various water harvesting technique. The efficient way for sub-surface water storage, soil moisture conservation or ground water recharge technologies should be adopted properly under water resource development plan.

The various measures adopted under soil and water harvesting is:

- Vegetative barriers.
- Building of contour bunds along contours for erosion.
- Furrow/Ridges and Furrow ridge method of cultivation across the slope.
- Irrigation water management through drip and sprinkler methods.
- Planting of horticultural contour species on bunds.

Components of Watershed Management

Land Management

The land management refers to keep all those properties of land in proper order, which likely to affect the soil yield potency. The land characteristics such as terrain, slope, formation, depth, texture, moisture, in-filtration rate and soil capability are the main to consider under land management activities for watershed development.

In broad sense the land management interventions includes following activities:

- Vegetative measures
- Structural measures
- Production measures
- Protection measures

The vegetative measures are the primary land management measures. The development of grass lands/pasture lands for erosion/soil loss control; adoption of contour farming and strip cropping practices on hill faces; growing of vegetations on barren lands or simply keeping the land under vegetations are the common practices used under vegetative measures for land management.

These measures are very effective to check the soil erosion, along with less cost expensive and easily practicable for the farmers. The practices such as development of vegetative cover, plant cover, mulching, vegetative hedges, grassland management, agro- forestry etc., are also included under this kind of measures.

The structural measures (mechanical conservation measures) such as bunding, terracing, check dams etc., are used at the steep lands for controlling the soil loss, especially when vegetative measures are ineffective. These measures are not so common as the vegetative measures because of involvement of heavy expenditure of money.

Similarly, the spurs and gabions used for stream bank erosion control; the gully plugging structures like drop structures, spillways etc., used for gully control; and farm ponds used for safe water storage in the farmland area, are also considered as mechanical measures for land management.

These structures offer their immediate effect on soil erosion/soil loss check, but very cost expensive, requires proper site selection, design and construction. Because of this reason, their construction is not possible by the farmers; the government normally executes it.

The production measures for land management include the practices such as mixed cropping, strip cropping, cover cropping, crop rotations, cultivation of shrubs and herbs, contour cultivation, conservation tillage, land leveling, use of improved variety seeds, horticultural practices etc. The objective of these measures is to enhance the production potential of the land either by conserving the soil or enriching the nutrient status.

The protective measures are the landslide control structures, gully plugging structures, runoff collection structures etc. Adoption of these measures depends very much on the land characteristics.

Water Management under Watershed Management

Under watershed management task the water management is one of the very important components. A good water potential in watershed provides a conducive path for its overall development. In watershed the main source of water is the rainfall; however, the incoming ground water from surrounding areas also shares to some extent. A large portion of rainwater is lost either due to flowing away (runoff) from the area or by some other means.

In order to manage the rainwater, it is very essential to check the out flowing rain water. It could be done by constructing the structures like pond, reservoirs etc. in the area. Also, the rain dependent farming systems can be practiced for better utilization of rainwater is also considered as a measure for water management. Apart from conserving the rainwater, their judicious use either for crop production or other farm operations, also play very significant role in water management.

As for as the water management regarding irrigation point of view is concerned, the selection of most suitable irrigation method depending on the crop, soil, land topography, availability of water in the area etc., is very important. Those irrigation methods should always be at priority, which have better water use efficiency, lesser loss of water etc. Similarly, the choice on cropping system, crop variety, and crop duration etc., based on the water availability can also be very effective in water management.

Overall, various interventions followed for water management are outlined below:

- Rainwater harvesting.

- Ground water recharge.

- Maintenance of water balance.

- Preventing water pollution.

- Economic use of water.

In watershed the water conservation by rainwater harvesting is most significant as compared to the other means. The harvested rainwater can be retained for the duration of its need by designing and constructing the suitable structures in light of the same.

The rainwater harvesting can be in the form of profile water conservation or surface water storage. The water conserved in the topsoil profile is the profile water conservation. Using the practices of tillage operations such as conservation tillage, zero tillage, mulch tillage etc. it can be achieved.

Depending on the moisture content in the topsoil profile a suitable crop can be taken successfully. Also, if the quantum of rainwater is very high then a part of that gets percolate to the lower soil profile and joins to the water table. This happening is called ground water recharge. There have been formulated several water harvesting techniques, worldwide.

However, few simple and cost effective rainwater-harvesting structures are listed as under:
- Percolation pits/tanks
- Farm ponds
- Bunds and terraces
- Reservoirs
- Community tanks
- Water spreading

Biomass Management

In a watershed the task of biomass management can be achieved by following intervention areas:
- Eco-preservation,
- Biomass regeneration,
- Forest management and conservation,
- Plant protection and development of social forestry,
- Increasing productivity of animals,
- Income and employment generation activities,
- Coordination of health and sanitation programmes,
- Better standard of living of people,
- Eco-friendly life style of people,
- Formation of learning community.

Watershed Management Practices

A watershed may involve a host of problems related to soil, water, society etc., and constraints in between. To remove all the associated problems mere should be need-based objectives. For achieving the target, various practices need to follow in the watershed.

Amongst them the most common practices are listed as under:

- Management purpose
- Increasing infiltration rate
- Increasing water holding capacity of soil
- Preventing soil erosion
- Method and accomplishment

The practices to be used for watershed management should be decided on the basis of the management objectives, exactly. Otherwise, the expected result may not be possible to get achieve. For example – if there is need of enhancing the ground water potential in the watershed then priority must be given for those practices, which augment the ground water recharge, effectively. A little deflection in selection of practices may cause in-conducive result.

The task of increasing infiltration rate is to enhance the soil moisture status; and accordingly to grow the crop depending on available soil moisture. On the other hand, the enhancement in infiltration rate causes reduction in level of runoff yield from the watershed, which in turn to affect the net available water for reservoir storage or satisfying the demand of other's need. The practices to enhance the infiltration rate are the tillage practices, cropping system, addition of organic materials etc.

The tillage practices make the topsoil surface in loose condition as result when rainfall takes place then a large amount of that gets enter the soil. The crops enhance the infiltration rate in significant way. In cropped field there is development of surface roughness, which obstruct the overland flow. Because of this reason the rainwater gets more time to retain over the land surface, as result a huge amount of water gets infiltrate into the soil. The organic matter in the soil improves the soil texture, favorable to enhance the infiltration rate.

The status of water holding capacity of the soil falling in the watershed plays very important role to develop overall effects on watershed behavior, either that is in respect of increasing the runoff amount, soil erosion/soil loss or making the soil properties better for good crop yield. The water holding capacity of soil can be improved by adding organic matters in the soil.

The management of watershed with severe soil erosion problem requires very attentive measures to check the erosion, immediately. The practices to be used for erosion control depends very much on the erosion intensity, soil features, cropping practices, mainly. If erosion intensity is not very high and land slope is in mild range then agronomical measures could be significant to check the soil erosion. On the other hand, when erosion rate and slope steepness of the area is very high then agronomical measures are not effective to check the erosion.

For this situation, the mechanical measures such as bunding, terracing etc., are very effective to check the soil erosion. Similarly, for gully erosion control a host of practices/methods have been devised such as drop structures, check dams, gabions etc., can be suitably used for watershed management.

In watershed, if there is stream bank erosion problem, then it can be tackled by using various methods such as spurs, gabions and agronomical measures depending on the bank situation and stream flow rate. The suitable measures for erosion control have been enlisted in Table.

Table: Various erosin/soil loss control practices/measures followed under watershed management.

S. No.	Practices/Measures
1.	Agronomical measures • Strip cropping • Pastures farming • Grass land farming • Wood lands

2.	Engineering measures (Structural practices)
	
	 • Contour bunding
	
	 • Terracing
	
	 • Construction of earthen embankment
	
	 • Construction of check dams
	
	 • Construction of farm ponds
	
	 • Construction of diversion
	
	 • Gully control structure
	
	   i. Rock dam
	
	   ii. Establishment of permanent grasses and vegetations
	
	 • Providing vegetative and stone barriers
	
	 • Construction of silt dentention tanks
	
	 • Construction of spur for stream bank erosion control

The watershed management measures can be grouped under following two main categories:

1. In terms of purpose; and

2. Method and accomplishment.

In first category of management measures, the land use and treatment measures are considered which are effective to increase the infiltration rate and water holding capacity of the soil and also prevent the soil erosion from watershed. Under this group, all the biological and mechanical methods employed for erosion control, are included.

In second category those measures are included, which are planned primarily for the management of water flow. The flood water retarding structures, stream/channel improvement to make their carrying capacity sufficient, minor flood-ways, sediment detention in watershed etc., are counted for watershed management work.

Chapter 6
Hydrology of Forests and Wetlands

Wetland hydrology deals with the study of the flow of water into and out of a wetland along with its interaction with other site factors. The branch of study which seeks to analyze the distribution, storage and quality of water along with the hydrological processes in forest-dominated ecosystems is called forest hydrology. This chapter discusses in detail the hydrology of wetlands and forests as well as the different types of forests.

Wetland Hydrology

Wetlands are areas where water covers the soil, or is present either at or near the surface of the soil all year or for varying periods of time during the year, including during the growing season. Water saturation (hydrology) largely determines how the soil develops and the types of plant and animal communities living in and on the soil. Wetlands may support both aquatic and terrestrial species. The prolonged presence of water creates conditions that favor the growth of specially adapted plants (hydrophytes) and promote the development of characteristic wetland (hydric) soils.

Categories of Wetlands

Wetlands vary widely because of regional and local differences in soils, topography, climate, hydrology, water chemistry, vegetation and other factors, including human disturbance. Indeed, wetlands are found from the tundra to the tropics and on every continent except Antarctica. Two general categories of wetlands are recognized: coastal or tidal wetlands and inland or non-tidal wetlands.

Coastal/Tidal Wetlands

Coastal/tidal wetlands in the United States, as their name suggests, are found along the Atlantic, Pacific, Alaskan and Gulf coasts. They are closely linked to our nation's estuaries where sea water mixes with fresh water to form an environment of varying salinities. The salt water and the fluctuating water levels (due to tidal action) combine to create a rather difficult environment for most plants. Consequently, many shallow coastal areas are unvegetated mud flats or sand flats. Some plants, however, have successfully adapted to this environment. Certain grasses and grass-like plants that adapt to the saline conditions form the tidal salt marshes that are found along the Atlantic, Gulf, and Pacific coasts. Mangrove swamps, with salt-loving shrubs or trees, are common in tropical climates, such as in southern Florida and Puerto Rico. Some tidal freshwater wetlands form beyond the upper edges of tidal salt marshes where the influence of salt water ends.

Inland/Non-tidal Wetlands

Inland/non-tidal wetlands are most common on floodplains along rivers and streams (riparian wetlands), in isolated depressions surrounded by dry land (for example, playas, basins and "potholes"), along the margins of lakes and ponds, and in other low-lying areas where the groundwater intercepts the soil surface or where precipitation sufficiently saturates the soil (vernal pools and bogs). Inland wetlands include marshes and wet meadows dominated by herbaceous plants, swamps dominated by shrubs, and wooded swamps dominated by trees. Certain types of inland wetlands are common to particular regions of the country.

Many of these wetlands are seasonal (they are dry one or more seasons every year), and, particularly in the arid and semiarid West, may be wet only periodically. The quantity of water present and the timing of its presence in part determine the functions of a wetland and its role in the environment. Even wetlands that appear dry at times for significant parts of the year - such as vernal pools-often provide critical habitat for wildlife adapted to breeding exclusively in these areas.

Geomorphic Position

Wetlands are a fundamental hydrologic landscape unit that generally form on flat areas or shallow slopes, where perennial water lies at or near the land surface, either above or below. Wetlands tend to form where surface and ground water accumulate within topographic depressions, such as along flood plains, within kettles, potholes, bogs, fens, lime sinks, pocosins, Carolina Bays, vernal pools, ciénegas, pantanos, tenajas, and playas, and behind dunes, levees, and glacial moraines. Seepage wetlands form where ground water discharges on slopes, as well as near the shores of streams, lakes, and oceans. Fringe wetlands also form along shorelines, with periodic inundation not caused by ground water discharges but, rather, by water exchanges with adjacent waterbodies, such as by periodic floods and tidal action. And, finally, perched wetlands form above low-permeability substrates where infiltration is restricted, such as above permafrost, clay, or rocks.

Brinson provides a methodology for using hydrogeomorphic indicators to classify wetlands based on their unique hydrologic, geomorphic, and hydrodynamic characteristics. In this way, the dominant landscape and hydrologic factors can be synthesized to better develop an understanding of wetland forms and functions.

Energy as the Driving Force

The direction and rate of water movement into and out of wetlands is controlled by the spatial and temporal variability of energy. A change in energy with distance generates a force that causes water to move from zones of high energy to zones of lower energy. Gravitational forces account for most water movement, in that water tends to flow from higher to lower elevations. Resisting the gravitational force are viscous (friction) forces that retard the fluid velocity. Inertial (momentum) forces resist a change in velocity, causing water to move at a constant velocity and in a straight line, unless additional energy is expended to either accelerate, decelerate, or deflect the water.

Water can also move due to a change in pressure, from zones of high pressure to zones of low pressure. This is common in ground water systems, where confined aquifers flow to the surface because of the greater pressure at depth. Artesian flow from a confined aquifer to the surface occurs when the recharge area to the aquifer lies at a higher elevation than the ground surface where the discharge occurs. Classical artesian springs exist in lowlying areas that are supplied with flows from higher elevation areas.

Wetlands are normally found in low-energy environments—that is, in areas where water normally flows with a slow velocity. This results, in part, because the land surface is relatively flat in these areas. Because wetlands lie in relatively flat landscapes, their surface area expands and contracts as the water stage changes, allowing for the storage of large volumes of water. Wetlands therefore serve as a moderator of hydrologic variability—storing flood flows and reducing flow velocities during wet weather in particular. In addition, shallow depths and low slopes, consistent with low energy environments, are important for trapping nutrients and sediments.

Hydrologic Measures

Three hydrologic variables can be defined that are useful for characterizing wetland hydrologic behavior; the water level, hydropattern, and residence time. Each of these wetland descriptors are described in greater detail in subsequent sections. What follows here is a brief introduction of these concepts.

One hydrologic descriptor is the general elevation of wetland water levels relative to the soil surface. Open water usually occurs in deeper areas with few, if any, emergent macrophytes. Any vegetation present in these areas is usually not attached to the wetland bottom, but vegetation may be floating on the water surface. An emergent zone may also be present in areas shallower than the open water zone, containing substantial quantities of emergent macrophytic vegetation, either living or dead. Yet, other wetlands may have large areas of exposed, saturated soil that is generally covered with macrophytic vegetation. The water level can, therefore, be used as an indicator of the vegetation types likely to occur in each of these zones.

A second descriptor of wetland hydrology is the temporal variability of water levels. The timing, duration, and distribution of wetland water levels are, together, commonly referred to as the wetland hydropattern, which incorporates the duration and frequency of water level perturbations. The hydropattern of some systems, such as tidal marshes, fluctuate dramatically over short periods of time; other systems, such as seasonally flooded bottomland hardwood communities, fluctuate more slowly over time. Yet, other wetland systems are more static and may not display substantial short- or long-term variability. The wetland hydropattern is a function of the net difference between inflows and outflows from the atmosphere, ground water, and surface water.

A third descriptor of wetland hydrology is the residence, or travel time, of water movement through the wetland. Some wetland systems exchange water quickly, with water remaining within the wetland for only a short duration of time, while water may travel very slowly through other wetland systems. The residence time is the ratio of the volume of water within the wetland to the rate of flow through the wetland. Short residence times occur when the flow through the wetland is large compared to its volume—longer residence times occur when the flow is small compared to its volume. The residence time of a wetland is often related to its hydropattern in that wetlands with large water level fluctuations may have shorter residence times, such as in tidal marshes. On the other hand, some wetlands may fluctuate rapidly due to large changes in inflow, yet have very long residence times due to slow loss rates.

Wetland Water Level

An important feature of wetlands is the condition of oxygen deficiency in wetland soils. Anaerobic conditions develop more quickly in saturated soils than in unsaturated soils due to low oxygen solubility in water, slow rates of water advection, and slow diffusion rates of oxygen through water. Anaerobic conditions in wetland soils affect vegetation by creating adverse conditions for root survival and growth. Thus, the presence of water substantially affects soil oxygen concentrations, which affects plant growth and survival.

Yet, despite these low oxygen concentrations, wetlands are among the most biologically productive ecosystems on the landscape. They support a diverse assemblage of vegetative species having special physiological adaptations that enable them to survive and prosper in these otherwise harsh growing conditions. Many biogeochemical reactions occur within these low oxygen zones, as noted elsewhere in this document.

Tidal Wetland Types

- Subtidal: Tidal water permanently covers the land surface.

- Irregularly Exposed: Tidal water usually covers the land surface, but is not exposed daily.

- Regularly Flooded: Tidal water alternately covers and daily exposes the land surface.

- Irregularly Flooded: Tidal water covers the land surface less often than daily.

Non-tidal Wetland Types

- Permanently Flooded: Water covers the land surface throughout the year in all years. Vegetation is composed of obligate hydrophytes.

- Intermittently Exposed: Water covers the land surface throughout the year except in years of extreme drought.

- Semipermanently Flooded: Water covers the land surface throughout the growing season in many years. The water table is at or very near the surface when the land surface is exposed.

- Seasonally Flooded: Water covers the land surface for extended periods, especially early in the growing season, but is absent by the end of the season in most years. The water table is at or near the surface when the land surface is exposed. Saturated water never covers the land surface but the soil is saturated to the surface for extended periods during the growing season.

- Temporarily Flooded: Water covers the land surface for brief periods during the growing season but the water table usually lies well below the surface for most of the season. Plants that grow both in uplands and wetlands are present.

- Intermittently Flooded: Water covers the land surface for variable periods with no detectible seasonal periodicity. Long periods of time separate periods of inundation. The dominant plant com munities under this regime may change as soil moisture conditions change. Some areas may not exhibit hydric soils or support hydrophytes.

- Artificially Flooded: The amount and duration of flooding is controlled by means of pumps or siphons in combination with dikes or dams.

Water levels in wetlands serve as an indicator of the dissolved oxygen state of the soilwater system. Wetter systems generally have higher water levels and lower soil dissolved oxygen concentrations, while drier systems have lower water levels and higher dissolved oxygen concentration. A general relationship between wetland water levels and hydric states is shown in table. Note that a distinction is made between soil saturation and soil surface inundation. Some systems may be flooded for part of the year and still have low pores water soil saturation, and vice versa. Low soil dissolved oxygen concentrations and reducing conditions may result in both cases.

Wetland water levels (also called the stage) may not be indicative of soil saturation. The zone of pore saturation may extend above the water table due to capillary rise in fine grained materials. Capillary rise results from the natural tendency of water to adhere to soil surfaces and other water molecules. Because the capillary height of rise can extend for several meters above the water table in fine-grained materials, the soil may be entirely saturated even when water levels are below the ground surface.

Hydrographs

A hydrograph relates the stage, or water level, as a function of time. Between storms, water levels in wetlands normally decline slowly over time, rising in response to precipitation. The rising limb of the hydrograph corresponds to the period from when water levels begin to rise following a precipitation event. The peak stage corresponds to the time when water levels reach their highest level. The falling limb of the hydrograph corresponds to the period following the peak and lasts until the next storm.

The time to peak is the length of time between the peak precipitation and peak stage. Times to peak are short in urban areas with large impervious surfaces and channels that have been modified to increase stream velocities. Times to peak are longer in forested areas with few impervious surfaces and channels with many obstructions that slow the passage of water. Another term, the time of concentration, is the time required for flow to travel from the most distant point on the watershed, and is a function of the same factors that affect the time to peak.

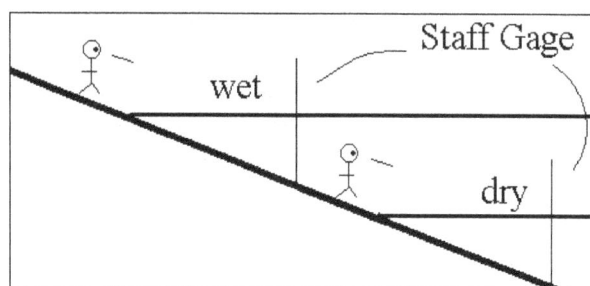

Figure presents that Multiple staff gages can be used to determine wetland water levels for a wetland with highly variable stages. The higher gage is used during wet periods and is easier to read than the more distant one. Once water levels fall below the higher gage, then the lower gage is used.

Monitoring Water Levels

Water levels in wetlands can be determined using a staff gage if the water level is above the ground surface. A staff gage is a vertical scale that serves to indicate the elevation of water, or stage, with respect to a reference elevation. The staff gage is an inexpensive tool that should place in the wetland such that the base of the staff gage is always submersed in water. For ease of measurement, multiple staff gages can be placed at different depths, such that the nearest one is visible from the shoreline during wet weather. The deeper staff gage is used once water levels fall below the nearer gage. In this way, one submerged staff gage is always visible from the shoreline as water levels rise and fall.

In some wetlands, a water line can be observed on periodically submerged vegetation. The water line can indicate a high-water level after a flood. In these cases, floating debris— such as leaves, trash, or branches—is lodged in the canopy of trees or bushes. In other cases, the natural water level can be observed as a horizontal line on the sides of trees. This line typically represents the normal water level in the wetland. This line would not be visible during wet weather, but is more likely to be observed during drier periods.

If water levels are below the ground surface, then piezometers can be used to find the water surface. A piezometer is a small-diameter perforated tube that is installed within the soil at a specified depth. The perforated zone should be narrow to minimize interference between layers, and placed within a unique hydrogeologic unit such as a soil horizon or geologic layer.

Water levels can be inexpensively determined by lowering a weighted, chalk-covered steel measuring tape into the piezometer. The tape is lowered until at least one part of the tape is wet. The reading on the tape where the chalk has been wetted is subtracted from the reading taken on the tape at the top of the piezometer. A slightly more expensive technique is to use a depth-to-water detector, which provides an audio or visual signal when the water level is encountered. Another option is to use an automated water level recorder, such as a float or pressure transducer. The advantage of automated techniques is their suitability for conditions when water levels are both above and below the ground surface.

When water levels are below the ground surface, the degree of soil saturation can be measured using Time-Domain Reflectometry (TDR). TDR determines the water content using the electromagnetic properties of a wave pulse passing through a conducting set of rods (such as 3-mm stainless-steel welding rods) placed in the soil. TDRs provide the soil water content without the need for calibration. Because the air porosity is not measured using TDRs, additional measurements of the total porosity are required, generally by collecting soil cores. The soil saturation is the ratio of the water content to the total porosity.

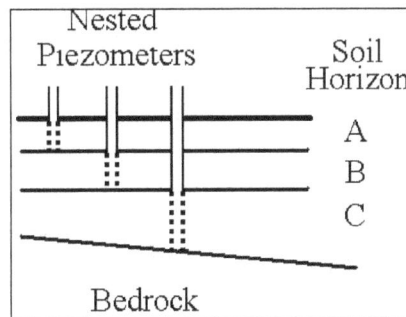

Piezometers are used to monitor water level changes in the subsurface. Multiple piezometers (called a nest) can be installed next to each other, one in each soil horizon, if vertical flow between soil horizons is expected. Additional piezometers can be installed in the bedrock if such materials affect the hydrology of the wetland.

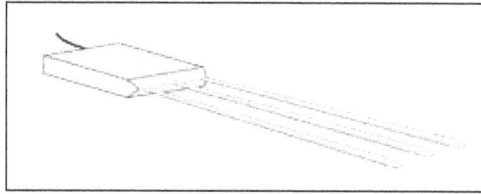

Time Domain Reflectometer probes can be used to monitor soil moisture saturation over time, providing an estimate of the water change in the subsurface.

The principle of TDRs is that the velocity of the electromagnetic wave along a conductor is a function of the dielectric coefficient of the media around the conductor. Larger dielectric constants cause slower wave velocities and, hence, longer travel times. Liquid water has a dielectric constant of 80.2 at 20 °C, while ice is 3.2, petroleum is 1.8 to 2.2, quartz is 4.3, and air is only 1.00. It is clear, therefore, that the wave velocity is substantially retarded as the water content of a soil increases.

Hydropattern

The temporal variability of water levels in wetlands results from dynamic changes in hydrologic inputs and outputs, and temporal changes associated with hydraulic controls within the wetland. Temporal changes in water level are important determinants for many aquatic flora and fauna. The reproductive success of these wetland species can be adversely affected when fluctuations are not correctly synchronized with their developmental stages.

The hydropattern is a distinctive feature of the hydrologic variability that describes the variation of water levels over time and space. Hydropattern is a recent term that is used to expand the traditional concept of hydroperiod (i.e., the frequency and duration of time that the wetland is saturated) by incorporating additional information about the aerial extent and timing of inundation. The aerial extent is important, especially for large, complex wetlands that contain a variety of wetland features.

Several approaches can be used to characterize temporal changes in wetland stage (i.e., water levels). The easiest approach is to plot wetland stage as a function of time (called the hydrograph). The hydrograph shows the stage for a period of time that captures the range of possible hydrologic variability. Inter-annual, seasonal, event, and daily water level fluctuations may become apparent using such an approach.

Using the observed water levels, a plot of flooding duration versus wetland stage can be constructed. This plot provides a descriptive summary that indicates how long a typical flood occurs for each stage. Lower elevations have longer durations of flooding than higher elevations. This approach is useful for characterizing water level variability by generating a stage-duration relationship that quantifies the duration in time that a specified water level is exceeded. In this case, the period of time that water levels exceed the specified stage (or range in stages) is described. This approach should also consider the seasonal nature of inundations by dividing the data into specific time frames.

While the stage-duration approach successfully captures the duration of time that the system is flooded, it fails to characterize the frequency with which this occurs. That is, the number of times that a water level exceeds a specified stage for a specific time period is not quantified. An alternative approach is to quantify the frequency in time that the wetland is observed to exceed a range of specified stages. This approach yields a cumulative frequency table or plot that can be used to calculate exceedance probabilities. The mean, median, and extreme stages (e.g., 1, 10, 50, 90, and 99 percentile probabilities) can be estimated using the exceedance probability plot.

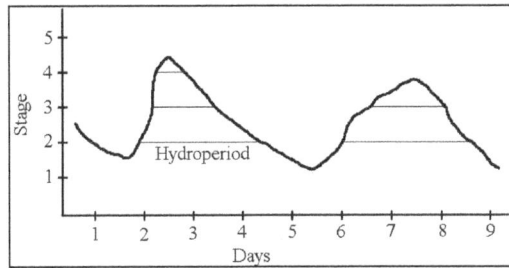

Hydrograph for a short period of time showing the water level variation.
Note that the hydroperiod is marked for a few stages.

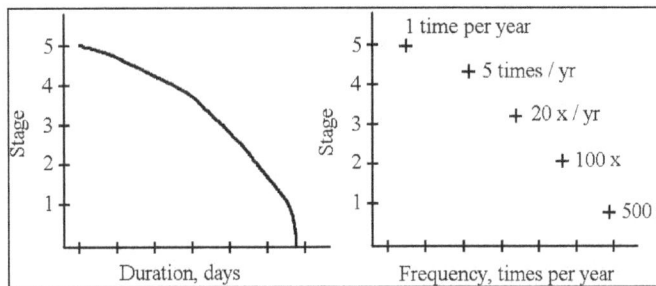

Left: Wetland hydroperiod plot showing duration of flooding versus stage. Right: Wetland stage-frequency plot showing number of exceedances per year. Note that the longest duration of flooding occurs at the lower stages, and vice versa. Lower stages have a higher frequency of being flooded than higher stages.

A significant drawback using the exceedance probability approach is that the correlation structure between individual observations may or may not be captured. That is, a system which water levels vary slowly over time can display the same frequency distribution of water levels as does a system that varies quickly. In effect, the amplitudes of the fluctuations are described, but not the duration. An additional problem is that the frequency diagram, in aggregate, may poorly convey daily and seasonal behaviors. Partitioning or stratifying data sets into seasonal or other periods may improve the characterization of water level conditions.

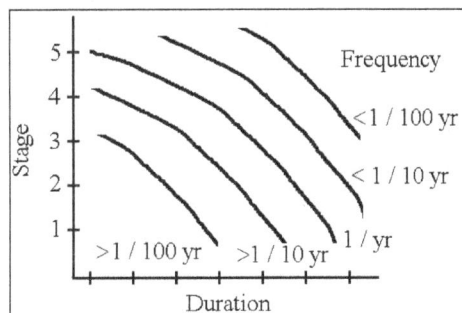

Wetland stage-frequency-duration plot showing duration of flooding versus stage
for a range of frequencies.

To overcome these limitations, a stage duration-frequency (SDF) curve can be constructed. The SDF plot is analogous to intensity-duration-frequency (IDF) curves used in precipitation analysis.

SDF curves indicate the frequency that a depth-duration relationship is observed. For large, complex wetlands the hydrodynamic behavior may be very different from one area to the next. Characterizing the hydropattern for the entire wetland is a substantially more difficult exercise than for a small, uniform wetland.

Hydrologic Residence Time

The hydrologic residence time is used to evaluate the time required for a hydrologic input to pass through the wetland. The residence time, 't' for a system with constant volume and flow rate is simply the ratio of the volume of water within the wetland, V, to the flow rate, Q, or, 't' = V / Q.

The estimated residence time is only appropriate for conditions of: 1) piston-flow, such as a First-In-First-Out (FIFO) queue; 2) steady (i.e., constant) flow; 3) single locations of inflow and outflow; and, 4) no atmospheric or ground water exchanges. The estimated residence time is not appropriate for conditions when water within the wetland mixes, multiple inflows and/or exchanges occur at different points within the wetland, or flow into the wetland is not constant over time.

If these conditions are not met, then the above equation only provides an estimate of the average residence time—actual residence times now varying over time and space. Functions for describing the distribution of residence times may be found for simple systems. For example, the exponential function can be used to determine the residence time distribution for a fully mixed system with constant inputs over time.

Different parts of a wetland may exhibit different hydrologic residence times. Water in active, flow-through sections of a wetland may have shorter residence times than water in inactive, isolated parts of a wetland. While each section may have identical hydropatterns the flow is concentrated in one area, leaving other areas with stagnant conditions. The same equation can be applied regardless; in this case, each section would be characterized using the volume of water present in the section, and the flow rate would be characterized using the flow into the section of interest.

Residence times for dynamic systems are more difficult to calculate than those with a steady flow. In these cases, the residence time changes—with increasing residence times during periods when outflows exceed inflows, and decreasing when inflows exceed outflows. This is because adding so-called new water or removing old water from the system decreases the age.

While hydraulic residence times can be calculated using the above equation, tracer tests can also be conducted to confirm these calculations. Conservative tracers (i.e., nonreactive tracers that move passively with the water velocity) can be added at the inflow point of a wetland, and then tracer concentrations can be monitored at the outflow location. A breakthrough curve describes the resulting tracer concentration over time. The time required for the median concentration—when the outflow concentration equals half of the input concentration—provides an estimate of the average residence time of the system.

Wetland Water Budgets

Wetland water levels, the hydropatterns and residence times are influenced and controlled by hydrologic inputs and outputs. In many cases, the wetland conditions observed are influenced, in large part, by the gains and losses of water. A water budget is used to account for the inputs and outputs to the wetland. The exchanges can be with the atmosphere, with ground or surface water, or by tidal action. The sum of all exchanges is what affects wetland water levels, i.e., if atmospheric exchanges cause an increase in water storage but ground water exchanges deplete these storages, then the total effect is the balance of the two.

The water balance equation summarizes this concept:

$$DQ = I = O = \frac{DV}{Dt}$$

where 'D'Q is the difference between inflows, I, and outflows, O, and the 'D'V term represents the change in water storage over a period of time, 'D't. This equation means that the volume in storage increases whenever the inflows exceed the outflows, and vice versa. Because water levels are directly related to the storage volume, an increase in storage volume always results in an increase in water levels:

$$\frac{Dh}{Dt} = \frac{DO}{A} = \frac{I-O}{A}$$

where 'D'V=ADh, Dh is the change in water level, and A is the wetland area. This relationship holds because the change in storage equals the product of the area of the water shed and the change in water level. This relationship becomes more complicated, as noted below, whenever the water level falls below the ground surface. In these cases, the mineral and organic soil materials release less water because of their porosity and ability to retain water.

Balancing Inflows with Outflows

Potential wetland inputs include precipitation directly onto the wetland, direct overland flow, surface water inputs from rivers, streams, and marine sources, overbank flow, and ground water sources including subsurface, lateral unsaturated and saturated flow from uplands to toe-slope and flat landscapes.

Balancing the inputs are the possible outputs, including evaporation, transpiration, ground water recharge, and surface water outflows. Water levels rise over time when hydrologic inputs exceed outputs, and fall when outputs exceed inputs. The change in water levels can be described using the following equation:

$$Dh = \frac{DV}{A} = \frac{DQ}{A} Dt$$

where Dh is the water level change, DV=DQ Dt is the net change in the input water volume to the wetland, DQ=I-O is the net change in inputs less outputs, Dt is the time step, and A is the wetland storage area, equal to the volume of water released per unit change in water level.

Hydrologic inputs and outputs are expressed in units of volume or depth per unit time. Water levels, however, incorporate no explicit unit of time—only water level changes are expressed in terms of units of depth per unit time. Thus, observed water levels are the result of accumulated water level changes over time and can be expressed as:

$$h(t_i) = \sum_{j=0}^{\cdot} Dh(t_i - t_j) = \sum_{j=0}^{\cdot} \frac{DQ(t_i - t_j)}{A_e} Dt$$

where Dt = $t_i - t_j$. This integration of water level changes means that wetlands reduce water level fluctuations by storing water during wet periods and releasing it during dry periods. The fact that wetland water levels are accumulators of hydrologic change over time and space makes them sensitive to even small changes in environmental conditions. That is, even small alterations can manifest themselves as large changes in wetland conditions when accumulated over space and time.

In addition to a water balance equation, the mass of dissolved and suspended matter carried by the water can be balanced. The mass balance equation is written as:

$$\frac{DM}{Dt} = DL = L_1 - L_2$$

where M is the mass of dissolved or suspended matter carried by the water, 'D'M is the change in mass between two points, 'D' t is the time interval, DL is the change in load, L_1 is the inflow load, and L_2 is the outflow load. Clearly, the rate of change in mass per unit time is a function of the balance between inflows and outflows.

Most water quality measurements are not based upon a load assessment. Instead, the solute concentration is normally measured.

The relationship between the load, L, and the concentration, C, is found by noting that:

L = C Q

where Q is the flow rate. This is because the concentration is:

$$C = \frac{L}{Q} = \frac{M}{V}$$

or mass per unit flow rate, which is just the mass, M, per unit volume, V.

Stage-Area-Volume Relationships

Changes in water depth must normally be converted to changes in water volume. This conversion need arises because inflows and outflows are measured in terms of water volume, while water levels within the wetland are measured in terms of water depth. A conservation of mass approach can be used to equate the two quantities. The conversion from water depth, h, to water volume, V, requires knowledge of the effective storage area of the wetland, A:

$$A = \frac{DV}{Dh}$$

where 'D'V is the change volume of water and 'D' h is the change in water level.

For conditions when water levels are entirely above the ground surface, the water volume change per unit depth equals the wetland area. The effective storage area may change as the wetland grows in size during high stage, thus requiring the use of a table or plot of wetland stage versus area.

Subsurface Water Storage

An additional complication arises when wetland stages are below the ground surface. In this case, the specific yield (i.e., the drainable porosity) of the organic and mineral sediments must be known. The specific yield is the volume of water released per unit area of wetland per unit decline in water level. In many cases, organic and mineral sediments may remain at or near saturation as water levels fall. In these cases, only a small volume of water is released from the sediments as they drain.

Combining the storage above and below the ground surface yields the following expression for the effective storage area:

$$A = A_S + S_y A_e$$

where A_s is the area of submerged wetland, A_e is the area of exposed wetland, and S_y is the specific yield of organic and mineral benthic sediments, generally equal to the difference between the saturated water content and field capacity of the sediments.

Specific yields of sediments are strongly influenced by their particle size distribution and chemical composition. Sands have large specific yields, while clays and mineral soils have low specific yields. The specific yield can be determined by extracting core samples and determining their specific yield, or reference texts can also be consulted.

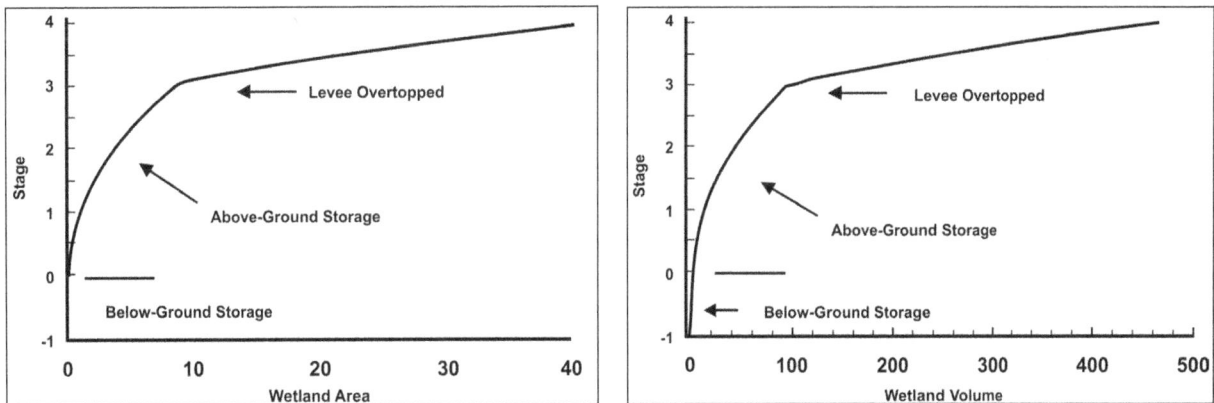

Figure presents Stage-Area (top) and Stage-Volume (bottom) curves showing changes in wetland area and volume as a function of stage. Flooded wetland area is zero once water stage drops below the ground surface in the deepest section of the wetland. Volume of water storage is not zero because the bed sediments may be able to store and release water. Wetland area also increases rapidly if a levee is overtopped.

The resulting areas are used to generate stage-area and stage-volume relationships from which changes in stage can be related to net changes in volume. In general, the stage volume relationship shows a sharp change in slope once water levels fall below the ground surface, as well as when water levels overtop a natural bank or levee. Synthetic stage-area and stage-volume curves are presented in Figure.

Determining Areas and Volumes

If sufficient detail is present, then wetland areas as a function of water elevation can be determined using topographic maps. Otherwise, surface mapping using a transit or level can provide cross-sections from which the volume and area can be determined. Alternatively, aerial photographs taken at different water stages can be used to estimate the stage area relationship.

One method for determining the volume of wetland is to add a known mass of tracer, allows the tracer to thoroughly mix in the wetland, and then measure the tracer concentration. Because the tracer concentration, C, is equal to the mass of tracer per unit volume of water, $C = M / V$, the volume of water within the wetland, V, is equal to the mass of tracer, M, divided by the tracer concentration, $V = M / C$.

If evapotranspiration is the only outflow and there are no inflows, then the tracer concentration increases over time as the volume of water within the wetland decreases. Likewise, if precipitation or surface and ground water inflows are present with no corresponding outflows, then the tracer concentration decreases over time, allowing the calculation of the wetland volume.

Water levels can be used in conjunction with the tracer data to obtain an estimate of the water balance. In effect, there are three mass balance relationships that can be used to estimate water budget components:

Water balance equation:

$$Q_1 - Q_2 = DV/Dt$$

where,

- Q_1 is the total inflows,
- Q_2 is the total outflows,
- DV is the change in storage volume, and
- Dt is the change in time.

Mass balance equation:

$$L_1 - L_2 = DM/Dt$$

where,

- L_1 is the total mass input,
- L_2 is the total mass output, and
- 'D'M is the change in mass.

Concentration equation:

$$C = M/V = L/Q$$

where,

- C is the concentration,
- M is the mass,
- V is the volume,
- L is the load, and
- Q is the discharge.

Water Budget Components

Water budgets are an important tool for characterizing the behavior of wetland systems. There are four general types of water sources and sinks in wetlands. The first includes atmospheric inputs and outputs, including rainfall, snow, evaporation, and transpiration. The second type of water exchange includes subsurface inflows and outflows. Another exchange mechanism results from interaction with surface water, including overland flow, as well as from rivers and streams. A final type of exchange occurs in marine systems that respond to tidal variations.

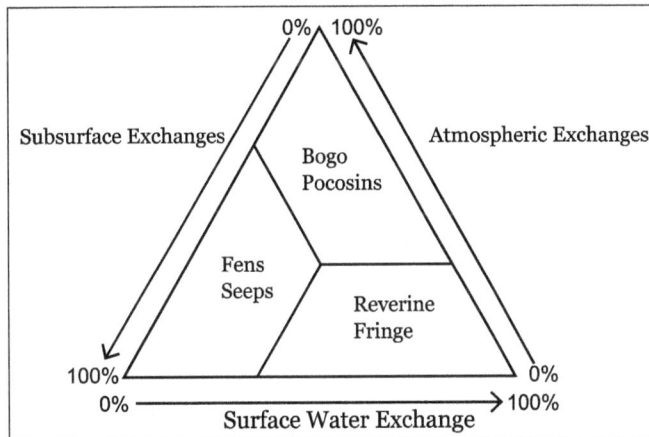

Relationship between hydrologic exchanges and nontidal wetland types

Figure presents a general characterization of wetland hydrologic exchanges for three of the four types of hydrologic exchanges (tidal wetlands are not included). The figure was adapted from Brinson, who originally characterized wetlands based on the type of hydrologic inflow. The figure has been modified to show that wetlands are affected by exchanges of water—both inflows and outflows. For example, atmospheric exchanges include evapotranspiration components as well as precipitation. Surface water exchanges incorporate releases from wetlands.

As noted by Brinson, wetlands have different types of inflow and outflow patterns. That is, some wetlands have simple exchanges with adjacent waterbodies, such as with a river during a flood such that the wetland receives water from the river during the rising stage of the flood and returns water to the river during the falling stage. Another example is the tidal exchanges along the coast, where the water moves in and out of wetlands to its original source.

In other cases, exchanges may be between different types of waterbodies and have a hybrid character. That is, inflows may be of one type (e.g., subsurface inputs) and outflows may be of another (e.g., evapotranspiration). Regardless, it is clear that identifying the key wetland hydrologic inflow and outflow components is a useful tool for understanding and managing wetlands.

Atmospheric Exchanges

Atmospheric exchanges, i.e., precipitation, evaporation, and transpiration, also need to be estimated for the water budget. Rain, snow, hail, sleet, freezing rain, fog drip, dew, and frost are various forms of precipitation resulting from the condensation of tropospheric water vapor. Water levels in wetlands that are dependent on atmospheric exchanges tend to be more affected by climatic signals than wetlands dependent on ground water sources. Lakes, like wetlands, tend to integrate climatic signals over time because of the longer residence time in these systems.

Precipitation generally occurs as a discrete event, characterized by using intensity, duration, frequency, and areal extent. In aggregate, precipitation events can be described using monthly and seasonal averages along with longer-term variability associated with climatic fluctuations. While regional precipitation networks can be used to estimate site conditions, the large spatial heterogeneity of precipitation patterns generally means that onsite precipitation measurements (either conventional or recording raingages) are needed when trying to obtain information for water balance analysis.

Table: Illustrative water budget components (mm/year) for selected wetland types.

	Atmospheric		Surface Water		Ground Water	
	Precip.	ET	Inflows	Outflows	Inflows	Outflows
Southeastern Swam p	1100	1600	10	30	100	500
Northern Bog	900	700	0	200	0	0
Floodplain	1000	1400	3000	2900	300	0
Prairie Pothole	600	400	0	0	100	300

The loss of water from a wetland by evaporation from the water surface, as well as by transpiration from plant leaf and stem surfaces, can have large effects on water levels. The combined processes are called evapotranspiration. Evaporation dominates when open water is present and vegetation is not. Saturated soils may lose nearly as much as open water, but not if a litter or mulch layer is present. Transpiration dominates in systems with little open water and large coverage of living vegetation. Evapotranspiration rates are affected by leaf and stem area, air, water, and plant temperatures, atmospheric humidity, wind speed, and the water potential of exposed soils.

Evapotranspiration losses from wetlands in close proximity are generally similar. This is because the source of energy for vaporization (i.e., the sun) is regionally uniform, and the availability of water for vaporization is similar owing to the lack of water limitations in wetlands. Forested wetlands may have greater evapotranspiration rates, however, due to higher leaf areas. Also, wetlands covered with dead vegetation may have lower evapotranspiration rates due to a lack of transpiration, a reduction in evaporation from shading, and poorer wind exchange. If increasing eutrophication leads to increased plant leaf area, then increased evapotranspiration water losses to the atmosphere could result.

Evaporation Theory

Whether water evaporates to—or condenses from—the atmosphere is a function of the energy state of water in the liquid and gaseous forms. The vapor pressure of water (a measure of the water content of the atmosphere) is the primary measure of the energy state. Dalton's law relates the rate of evaporation to the difference in vapor pressure between the air-water interface and the vapor pressure in the atmosphere at some distance from the interface:

$$E = c\left[e_i - e_a \right] = c\left[RH_i e_s \left(T_i \right) - RH_a e_s \left(T_a \right) \right]$$

where E is the evaporation rate, c is an evaporation rate coefficient, e_i is the vapor pressure at the air-water interface, e_a is the vapor pressure in the atmosphere at some distance away from the interface, and $e_s \left(T_a \right)$ is the saturation vapor pressure, which is a function of the air temperature, T_a.

The evaporation rate coefficient, c, is a function of wind speed and the type of evaporation surface, either soil or water. The vapor pressure in the air, $e_a = RH_a e_s \left(T_a \right)$ equals the product of the relative humidity of the air, RH_a, and the saturation vapor pressure, e_s, based on the air temperature, T_a.

The relationship between the saturated vapor pressure and temperature for the range of liquid water at standard atmospheric pressure is:

$$e_s = 6.11 \exp\left[\frac{17.3T}{T + 237.3} \right]$$

where e_s is the saturated vapor pressure in hPa (1 hPa = 1 millibar) and T is the temperature in °C.

The energy state of the liquid at the air-water interface is a function of the fluid potential, or pressure, at the interface. If a free surface is present, then the fluid pressure at the air-water interface is zero, and the water potential in the atmosphere just above the surface equals the saturated partial pressure of water within the atmosphere (termed the saturated vapor pressure). In this case, the relative humidity of the air just above the interface is equal to 100% (i.e., saturated with water vapor). If the water potential at the water surface is negative, due to osmotic potentials or negative pressures within soil pores or within plant stomata, then the relative humidity above the surface is no longer saturated. The equilibrium relative humidity, RH, as a function of fluid potential, 'y', and temperature, T, is:

$$RH = \exp\left[\frac{y}{RT} \right]$$

where R is the water vapor gas constant.

Monitoring Atmospheric Exchanges

Precipitation and evaporation can be readily measured using raingages and evaporation pans, respectively. These are relatively inexpensive and provide reliable estimates of daily atmospheric exchanges. A single raingage is usually sufficient for small wetlands (e.g., smaller than 100 ha), but multiple raingages may be required for larger wetlands, especially if significant spatial variation in rainfall is present.

Measured evaporation rates can be used to estimate evapotranspiration rates. A single evaporation pan is probably sufficient for all but the largest wetlands. While potential evapotranspiration derived from pan estimates (either manual or recording) can be used to estimate site conditions, the local effects of shading and wind shelter can adversely affect the accuracy of the measurements. Pan coefficients (the ratio of actual evapotranspiration to pan measurements) are reported to range from 0.54 to 5.3.

Automated raingages are available but more expensive than manual raingages. Automated evaporation pans are less reliable, and additional research is needed to improve their accuracy. If pan measurements are not available, then evaporation can be calculated using automated measurements of solar radiation, temperature, relative humidity, and wind speed.

Daily precipitation data should be plotted, along with daily evaporation. The difference between precipitation and evaporation can be compared to observed wetland water levels. In systems where atmospheric exchanges dominate the wetland hydrology, water levels rise during precipitation events and fall at a rate controlled by the evaporation rate.

Subsurface Exchanges

Subsurface inflows to wetlands (also called ground water discharge to wetlands) may result from shallow, topographically induced drainage from nearby uplands or from discharges of regional, confined aquifers. Subsurface outflows from wetlands (also called ground water recharge) may result from downward and lateral flow from the wetland to underlying surficial aquifers, and to deeper, confined aquifers where the con fining layer has been locally breached due to collapse or subsidence.

Shallow inflows may result from perched, or interflow, drainage on top of lower-permeability units within the unsaturated zone, such as clay beds, soil horizons, or even permafrost. Shallow subsurface inflows may also arise when the water surface within the wetland lies below the water table in the underlying surficial (unconfined) aquifer. In this case, the direction and magnitude of the hydraulic gradient can be estimated using aquifer water levels obtained in piezometers positioned in the vicinity of the wetland. Besides the hydraulic gradient, the water flow rate is also dependent on the permeability of the aquifer and any organic and mineral benthic sediments. The inflows may be concentrated at one or several points within the wetland. Inflows may also result from diffuse upward leakage, in which case the leakage is more uniformly distributed across the benthic materials.

Subsurface inflows from deeper sources may arise when confined aquifers discharge into the wetland. These discharges occur when the confining layer is breached due to subsidence or collapse, such as in karst areas. In this case, wetland water levels are controlled by the piezometric surface in the confined aquifer. In addition, confined aquifer discharges can occur when diffuse upward leakage moves through the confining layer into the overlying unconfined aquifer and from there to the wetland. Discharges from deeper sources are less likely to respond rapidly to individual storm events, tending to be more responsive to seasonal and longer-term changes.

Shallow inflows may respond more rapidly to individual storm events, as well as to seasonal and climatic changes. This is because interflow and water levels in shallow aquifers tend to be more sensitive to net changes in atmospheric flux (precipitation less evapotranspiration in nearby upland areas).

Shallow subsurface outflows may occur if the wetland is underlain by a layer of low permeability that allows the water to perch. In these cases, a low point on the perimeter of the wetland allows water to exit the wetland as either overland flow, channel flow, or interflow. These systems have water levels that are perched above the regional water table, and may also have an unsaturated (vadose) zone present between the water table and the perched wetland. Recharge to the underlying, surficial aquifer also occurs through the low permeability layer.

Downward movement of water is prevented in permanently frozen soils, i.e., permafrost, because any liquid water is converted to solid form by heat exchange in the underlying frozen unit. Yet, the ability of the underlying frozen unit to freeze water can be overloaded over time, resulting in a loss of the confining ability of the unit. The entire loss of the permafrost layer is possible if too much heat is added to the unit.

In some cases, wetland water levels are contiguous with those within the surficial aquifer. In these cases, flow through the surficial aquifer may be affected by the wetland. Normally, surficial aquifer water levels dip in the direction of water flow, while the water levels in the wetland are more horizontal. Thus, wetland water levels lie below the water table in the upgradient direction, while they lie above the water table in the down gradient direction. As a result, aquifer discharge conditions are present at the upstream end of the wetland, and aquifer recharge conditions are present at the downstream end of the wetland. This type of flow-through wetland may account for most of the flux of water through a wetland that has no readily apparent inflows or outflows.

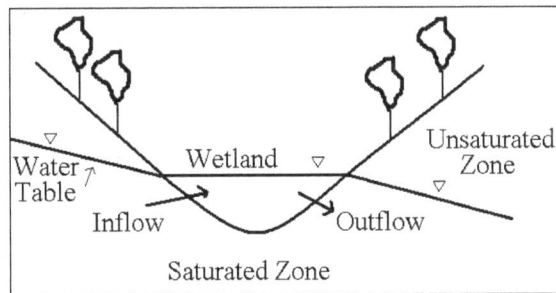

Effect of wetlands on surficial aquifer movement. Inflow to wetland is on side where water table levels are higher than wetland. Outflow is on side where water tables are lower.

Finally, recharge to deeper, confined aquifers may occur when subsidence or collapse has breached the confining layer that isolates the aquifer from the surface. In this case, water level increases in the wetland during wetweather periods may cause direct recharge to the deeper aquifer.

Ground Water Gradients

The gradient of ground water potentials governs the flow of water in the subsurface because the potential is a measure of the energy status of forces that cause water to move, and the direction of the forces controls this movement. The gradient is calculated using:

$$G = \left[G_x, G_y, G_z \right] = \left[\frac{Dh}{Dx}, \frac{Dh}{Dy}, \frac{Dh}{Dz} \right]$$

where G is the hydraulic gradient, composed of components in three directions, G_x, G_y, and G_z, which in turn are determined using the change in head, 'D' h, in each of the three directions, G_x, G_y, and G_z. Due to the layered nature of many geologic deposits, the hydraulic gradient can be simplified into horizontal and vertical components, with each layer having a unique horizontal flow pattern.

$$G_H = \left[G_x, G_y \right] = \left[\frac{Dh}{Dx}, \frac{Dh}{Dy} \right]$$

and

$$G_V = \left[G_z \right] = \frac{D_h}{D_z}$$

where 'G'$_H$ is the horizontal component and 'G'$_V$ is the vertical component of the hydraulic gradient. This approach is appropriate whenever horizontal layering is present. The magnitude of the horizontal component within each layer is found by determining the change in water levels with distance, while the vertical gradient between layers is found using the change in water level with depth between two adjacent layers.

The hydraulic gradient must be combined with the hydraulic conductivity to determine the ground water flux, or rate of volume flow. Like the hydraulic gradient, ground water flux can be separated into three components, two horizontal and one vertical:

$$q = \left[q_x, q_y, q_z \right] \quad qH = \left[q_x, q_y \right] \quad q_V = \left[q_z \right]$$

For flow through and across layered media, the horizontal component of flow, q_H, can be determined using the horizontal hydraulic conductivity, K_H, while the vertical component, q_V, uses the vertical conductivity, K_V:

$$q_H = -K_H G_H \quad q_V = -K_V \, G_V$$

These equations are the horizontal and vertical forms of Darcy's law. The negative sign indicates that flow is from regions of higher hydraulic head to regions where the head is lower. The estimated quantities are for flow at a point. The flows must be multiplied by the cross sectional area of the unit in question to estimate flow across the area.

The total flow, Q, is calculated using:

$$Q_H = A_H q_H \quad Q_V = A_V q_V$$

where A_H is the profile, or cross-sectional, area of flow, A_v, is the map-view area of flow, and Q_H and Q_v are the horizontal and vertical components of total flow, respectively.

Monitoring Subsurface Flows

Ground water gradients—and flows—normally vary over both space and time. Thus, a high resolution of temporal and spatial sampling is required to determine the flow field with any accuracy. This means taking multiple vertical and horizontal measurements at sufficiently frequent time intervals in order to capture any variability present in the system.

Monitoring ground water flows can be accomplished by placing an array of piezometers within and around the wetland. Multiple piezometers can be placed at different depths at each location to evaluate the magnitude of vertical flow. A nest of piezometers, composed of multiple piezometers placed at different depths, is needed whenever complex hydrostratigraphic conditions (e.g., layering) are present.

Piezometers are placed at multiple distances away from the wetland to determine the water table configuration in the neighborhood of the wetland. This network of piezometers, deployed at various locations and depths, is required to determine the three-dimensional characteristics of water potentials. Ideally, the locations of these water levels should form an equilateral triangle—not acute or obtuse triangles. Otherwise, the colinearity of the wells interferes with the estimation of the gradient.

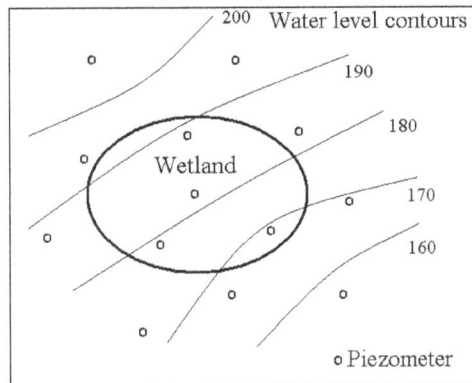

Network of piezometers required to map water levels in the vicinity of a wetland.
Note that the water level contours can be made based upon interpolation between measurements within individual piezometers.

Estimating the total flow of water into or out of the wetland requires an independent estimate of the hydrologic conductivity. Either aquifer tests can be conducted or standard tables of values can be used. Undisturbed core samples or field testing can also be used to estimate hydraulic conductivities. Care must be taken when using core data to estimate hydraulic conductivities in that the spatial variability of most geologic media is very high. In general, core samples tend to underestimate field hydraulic conductivities due to the difficulty in obtaining core samples for the very highest flow paths within the system.

Verry and Boulter report that the hydraulic conductivity of peat can be readily determined from its bulk density or unrubbed fiber content. They report that fibric peats have three-orders of magnitude higher horizontal hydraulic conductivity than sapric peats due to their larger pore sizes. Daniel notes a similar effect of decomposition on hydraulic conductivity.

Once the directional hydraulic conductivities and gradients have been found, then the total flow within the system can be calculated. There are often many uncertainties with this method, however, in that the spatial variability of both gradients and conductivities can be high.

Water quality sampling can be used as an independent method to determine the source of water within wetlands. For example, if shallow ground water is moving laterally into the wetland, and ground water is also moving upward into the wetland from a deeper aquifer, then the geochemical signature of each source can be used to evaluate the relative magnitude of each inflow relative to the total.

Surface Water Exchanges

Surface water exchanges with wetlands result from a large number of mechanisms, including over-land—or sheet—flow, direct exchange when the channel of a river or stream flows through the wetland, overbank flooding during wet weather when the channel is separated from the wetland by a levee or floodplain, and along the edges of lakes, estuaries, and the ocean. These surface water exchanges result in either constant or episodic hydrologic communication between the surface water and the wetland.

Streamflow can be divided into two types, baseflow and stormflow. Baseflow is that component of flow found during low flow periods, while stormflow refers to the response to precipitation events. If a stream was flowing before the rainfall (a typical situation), stormflow is the flow that occurs in addition to the baseflow that would have occurred if it had not rained.

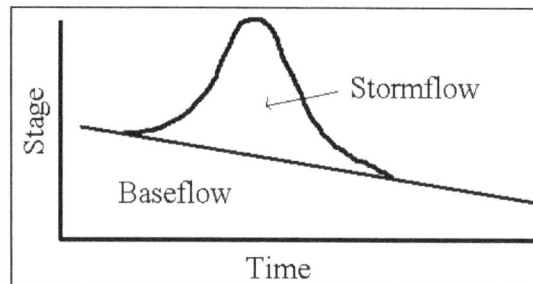

Water level behavior showing slow decline in baseflow along with
a discrete storm event with its associated stormflow.

Flood Attenuation

Not only can wetlands reduce flood velocities, they can also reduce flood wave velocities by decreasing water velocities at high discharges. Flood wave velocities travel at different rates than water velocities. The flood wave velocity, also called the flood celerity, c, is defined as the change in discharge, 'D'Q, per unit change in stream cross sectional area, 'D'A:

$$c = \frac{DQ}{DA} = \frac{D\left(\bar{v}\,A\right)}{DA} = \bar{v} + A\frac{D\bar{v}}{DA}$$

where Q=A v is the stream discharge, A is the stream cross-sectional area, and v is the mean stream velocity. This equation indicates that the flood wave velocity equals the water velocity plus a second term that is positive if the water velocity increases as the cross-sectional area (or, equivalently, stage) increases, and is negative if the water velocity slows as the area or stage increases. In other words, the wave velocity is faster than the water velocity if the water velocity increases with stage, and vice versa.

This concept can also be demonstrated using the ratio of the wave velocity to the water velocity, termed the kinematic ratio, k:

$$k = \frac{c}{v} = 1 + \frac{A}{v}\frac{Dv}{DA} = 1 + \frac{DInv}{DInA}$$

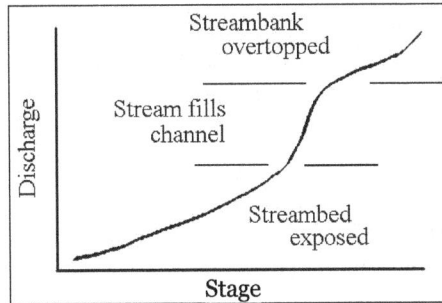

Rating curve showing relationship between water level stage and stream discharge. Note change in slope in relationship as different parts of the channel are wetted as the stage changes.

This relationship again illustrates the concept that the flood wave velocity is faster than the water velocity, k > 1, when the second term is positive, and vice versa. The second term in the equation is the critical parameter needed to control damaging flood waves. In effect, wetlands serve to sequester flood waters, thus slowing the average and incremental water velocities and flood wave travel times and reducing peak discharges. The flood wave velocity could actually be less than the water velocity if the velocity decreases with increasing depth.

Measuring Surface Flows

Surface water flows can be estimated using flow measurement control devices, such as weirs (which require a pool upstream and are not satisfactory when elevated sediment concentrations are present), flumes (which tend to flush sediments more effectively than weirs), and culverts (which are less accurate). The relationship between stage and streamflow discharge is called the rating curve. Once a rating curve has been developed, the stream discharge is readily found by observing the stage and then consulting the rating curve.

Open Channel Measurements

Stream discharge, Q, is obtained by establishing a stream cross section, and then measuring the stream velocity, v, across the section. Because the stream velocity and depth vary across the section, the total discharge is approximated by the sum of the discharge of subsections within the total section:

$$Q = \int A \; v \; dA^a \sum_{i=1}^{n} v_i A_i$$

where A is the cross-sectional area of the section, n is the total number of subsections, v_i is the average velocity in each subsection, and A_i is the area of each subsection, equal to the product of the width and average depth of the subsection.

For shallow streams, the average velocity is found at approximately 60% of the depth of the stream, measured from the surface down. For deeper streams, the average velocity is found by averaging two velocity measurements, at 20% and 80% of the depth.

Upstream and downstream variations in channel shape, as well as obstructions, may cause rapid changes in velocities within the cross-section. Thus, it is important to select a location where the channel appears to be of uniform width and depth and free of obstructions.

The site should also be selected so that backwater effects from downstream inflows are avoided. Another source of error occurs when the channel shape changes over time, so a solid bottom is preferred over a mobile bottom. Finally, a site located upstream of a knickpoint (a narrowing or shallowing of the river) is preferred over a site located downstream of a knickpoint. The knickpoint may cause subcritical (slow velocity) conditions upstream and supercritical (high-velocity) conditions (and even a possible hydraulic jump) downstream.

To construct the rating curve, the observed stream discharge is related to the river stage, measured using a staff gage. The staff gage is a vertical rule placed in a protected location. Repeated measurements of discharge over a range of stages is required.

Control Structures

Control structures avoid the vagaries of channel geometry by creating a uniform section. Table provides a summary of the various types of control structures along a brief summary of their attributes. A flume can be readily constructed with a uniform cross sectional area so that Q=Whv where W is the width of the flume, h is the depth of water in the flume, and v is the water velocity through the flume. In most cases, no unique relationship between depth and velocity can be established, being a function of the slope of the flume and the upstream and downstream conditions.

An improvement on the standard flume is to place a constriction in the flume (called the throat) that forces the flow to be subcritical (low velocity) upstream of the constriction, and supercritical (high velocity) downstream. Examples of this type include the Parshall Flume and the H-Type flume. The H-type flume was developed by the U.S. Department of Agriculture for measuring discharge in sediment-laden streams. Flumes require no upstream stilling basin and allow sediment to pass unimpaired through the structure. Ice, leaves, and other debris can still affect the reading, however.

Yet another control structure is the weir. A weir has a stilling basin upstream of a constriction (normally called the weir blade), and a free-fall below the constriction. The stilling basin is used to eliminate the velocity head, yielding H=z in the stilling basin, where z is the measured water surface elevation above the lowermost point on the weir blade.

Water flows out of the weir over the weir blade, which can take a variety of shapes, including triangular, rectangular, and trapezoidal. A submerged, circular orifice can also be used. The discharge is calculated using Q = vA = v (WH), where A=WH is the cross-sectional area perpendicular to flow above the weir blade, W is the width of the weir, and H is the depth of flow above the weir blade. So the general weir formula is:

$$Q = a\,H^{b}$$

where a accounts for the cross-sectional area as well as contraction and energy losses, H is the water surface elevation in the stilling basin above the weir blade, and b=0.5 for a flooded orifice, b=1.5 for a rectangular weir, and b=2.5 for a triangular weir. This equation holds for the broad-crested weir as well, with b=1.5.

Types of Channel Control Structures

Weirs:

- Stilling basin is located upstream of weir.

- Water level recorder is used to measure stage in stilling basin.

- Outlet structures include rectangular, triangular (v-notch), and Cipolletti (trapezoidal) shapes.

- Weir crests can be broad (flat lip) and sharp (knife-blade) crested.

- Flow is subcritical upstream of crest, supercritical downstream.

- Weirs collect sediments in the stilling basin, debris on weir blade.

Flumes:

- No stilling basin, only a narrow throat.

- Regular approach section.

- Passes sediment easily.

- Woody debris can be a problem.

Culverts:

- Four combinations of flow equations, flooded vs. Open upstream, flooded vs. Open downstream.

- Culvert should be a regular shape, round or rectangular, with no debris.

Weirs may not provide accurate estimates in several situations. One source of error occurs when the weir blade becomes blocked by ice or floating debris, such as leaves and branches. Another source of error arises when the weir basin fills with sediments, resulting in an inaccurate estimate of the total head. For all weirs, a staff gage or a water-level recorder is placed upstream of the constriction to measure the total head under subcritical conditions. Weirs tend to be more accurate than flumes but suffer from sediment accumulation in the stilling basin and debris obstructing the weir crest.

Culverts under roads can also be used as a control structure. Four types of flow conditions can be found for most culverts; upstream intake either submerged or open, and downstream discharge conditions either submerged or open. When the upstream end is flooded, and the downstream end is open, then the orifice solution for the weir equation can be used.

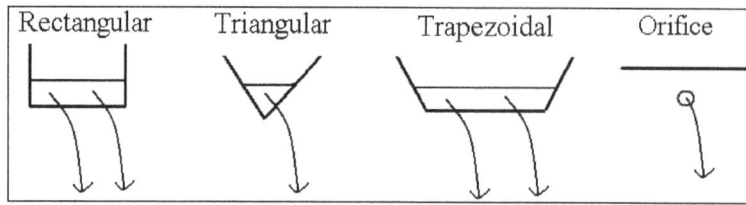

Three types of weirs (rectangular, triangular, trapezoidal, flooded orifice) used as control structures for measuring stream discharge. Water level (stage) is measured in weir basin upstream of weir blade.

When both upstream and downstream openings are submerged, then pipe flow conditions are present and discharge can be found using the difference in head between the two ends, the culvert length and diameter, and the type of culvert (smooth, corrugated, etc.). When the upstream end is open and the downstream end is either open or flooded, then the flow can be found using indirect techniques such as the Manning's equation.

Table: Two general types of weirs include broad-crested and sharp-crested. As the name implies, the broad-crested weir has a broad constriction in the direction of flow, while the sharp crested weir has a knife-edge blade that forms the constriction. A broad-crested weir consists of an outflow structure over which water flows for some distance before falling over the downstream edge. A sharp-crested weir is constructed so that the flowing water passes over a vertical, knife-edge, thus minimizing resistance with the weir blade.

Type	Weir Equation[‡]
Flooded Orifice	$Q = C\,A\,H^{0.5}$
Rectangular[§]	$Q = C_a\,W\,H^{1.5}$
Triangular	$Q = C_b\,tana\,H^{2.5}$
Trapezoidal	$Q = C_a\,W\,H^{1.5} + C_b\,tana\,H^{2.5}$
[†] Neglects contraction effects along weir blade edges	
[‡] H is elevation of water surface in stilling basin	
[§] Applies to broad-crested weirs as well	

Regardless of flow conditions, it is better if the culvert has a uniform shape throughout its length and is not obstructed with debris. Elevation measurements can be mademechanically using a water level recorder or visually using a staff gage.

Indirect Measurements

For situations in which control structures are not present and instream measurements not possible, Manning's equation is commonly employed to indirectly measure water velocity:

$$\bar{v} = \frac{1}{n} R^{2/3} S^{1/2}$$

where n is the Manning's roughness coefficient, R is the hydraulic radius, and S is the gradient in total head. The roughness coefficient is normally found in tables and is based on stream channel characteristics such as stream bed materials, amount of vegetation within the channel, variation in channel shape, and sinuosity. The hydraulic radius is defined as the ratio of the stream cross-sectional area to the wetted perimeter, R = A/P, which is approximately equal to the water depth in a shallow, wide channel. The total head gradient is the drop in total head per unit distance of stream channel.

Once the average water velocity, v, has been determined, the stream discharge, Q, can be estimated using:

$$Q = \bar{v}A$$

where A is the cross-sectional area of the channel.

Estimating Overbank Flows

Wetlands adjacent to riverine systems are often affected by overbank flows during stormflow periods. In these cases, the river spills out of its normal channel and overflows onto adjacent floodplains. The period of time that wetlands on the floodplains are affected by the duration of stormflows, and their amplitude. Instrumentation to monitor water levels in wetlands adjacent to riverine systems can be installed using techniques mentioned previously. Additionally, the U.S. Geological Survey provides estimates of overbank flooding frequencies for ungaged sites.

Tidal Exchanges

Coastal wetlands are similar in many ways to freshwater wetlands, except that they are transitional between marine and terrestrial environments. Coastal wetlands have unique attributes that distinguish them from both terrestrial and marine systems. It is, in fact, the combination of flooding and soils near the water surface that promotes ecologic diversity and productivity in coastal wetlands.

Coastal wetlands occupy similar landscape positions as lacustrine wetlands. Great variability exists, however, among coastal wetlands. Some coastal wetlands are dominated by the ebb and flow of the water levels of the ocean due to tides, termed tidal wetlands. The large magnitude of the daily tides, along with their regularity, result in unique wetland conditions. Other marine wetlands are more sheltered from tidal effects. Still others may be affected by water quality changes resulting from freshwater tributaries.

Tidal Effects

Many coastal wetlands function to dampen tidal and wave energies. Because viscous and turbulent drag through coastal wetlands dissipates energy, the magnitude of fluctuations generally diminishes with distance from the coast. The equation for a harmonic wave height with constant energy dissipation as a function of distance is:

$$h = h_o e^{-kx} \cos\left(wt - kx\right) \quad with \; k = \sqrt{\frac{w}{2D}}$$

where h is the wave or tidal height, h_o is the maximum magnitude of the fluctuation at the shoreline, 'k' is a damping coefficient that accounts for the dissipation of energy with distance, x is distance from the shoreline, w is the frequency of the fluctuation, t is time, and D is the hydraulic diffusion coefficient. The amplitude of the oscillation, $|h|=h_o\,e^{-kx}$, decreases with increasing frequency and distance from the shoreline, as does the lag, 'k' x. This means that low-frequency waves, such as daily and twice-daily tidal fluctuations, propagate deeper into coastal regions and are attenuated less and lagged more than high-frequency waves.

While the harmonic wave height equation may not be suitable for many coastal areas, it does combine most of the important factors (e.g., time, distance, wave frequency, and hydraulic resistance) that affect water and energy movement. An understanding of the local coastal morphology and vegetation is important, as are the energy inputs from marine sources and terrestrial surface water and ground water inflows. Together, these external factors influence how coastal wetlands are affected by hydrologic tidal exchanges.

Tides can affect wetlands many miles from any saline water sources. The hydrodynamic conditions imposed by a changing sea level cause rivers to flow more slowly during high tides, and vice versa. Thus, water levels in rivers upstream of the coast rise and fall in tandem with the tides. The magnitude of this effect is diminished with distance upstream and is a function of local channel features.

Other Inputs

Many coastal wetlands are affected by nearby freshwater inputs, especially in estuarine environments. In these cases, occasional, large stormwater inflows can cause rapid changes in the salinity, temperature, dissolved oxygen, and sediment concentration within the wetland. In some cases, these inputs can be beneficial, such as historical sediment deposition in the Mississippi River delta region of southern Louisiana—in contrast to the current practice of diverting stormflows away from coastal wetlands, which has led to regional subsidence and salt water intrusion. In other cases, these inputs can be detrimental, as when increased urban wastewater and stormwater inputs to coastal estuaries alter the natural conditions.

Ground water inputs to coastal wetlands may also be significant, and can take two forms, point and diffuse. Point discharges—such as springs—form when an underlying confining layer for an artesian aquifer is breached, allowing the upward flow of water. Diffuse upward leakage occurs when the confined artesian aquifer discharges over a large region.

In these cases, the leakage is more spatially uniform, with greater amounts of leakage occurring in low-elevation areas. Ground water exchanges may be more difficult to characterize than other sources, however. One promising method is the use of radioisotopes to differentiate between groundwater and other inputs.

Monitoring Coastal Wetlands

In some cases, distinguishing tidal from freshwater and ground water inputs can be achieved using geochemical information. The tidal water quality is clearly distinguished by its high concentrations of sodium-chloride type water, while freshwater inputs normally have markedly lower specific conductivities and total dissolved solids. Depending upon location, ground water inputs are intermediate, with possibly distinct geochemical signatures. For example, if carbonate aquifers are present, then a calcium signal may be present in these waters.

The degree of tidal flushing can thus be monitored using water quality data to characterize the residence times of water within the system. Estimates of fluxes can then be estimated based on the water balance equation.

Evolution and Alteration of Wetland Hydrology

Wetlands change over time. Natural processes such as sediments that fill wetlands and beaver activity, as well as accelerated processes such as upstream development and direct alteration of the wetland, all cause changes that affect wetland hydrologic behavior.

Natural Forces of Change

Wetlands are formed as the result of many geologic forces. Rivers form flood plains that provide a landscape position that enhances wetland development. Glaciers scour the landscape, leaving behind features that promote wetlands. Tectonic uplift and subsidence create depressional features that are favorable to wetland formation. Carbonate aquifers dissolve over time, leaving behind depressions where wetlands can form. Also, accelerated erosion transports sediment out of natural channels, leading to down-cutting and deepening of channels, which leads to a lowering of riparian water tables and the reduction of overland flows, both of which alter wetland saturation.

Wetlands can modify their environment as they mature. Peats may substantially modify the original landscape by filling in the depression they originally formed in. Other biological forces also promote wetland formation. Beaver create impoundments, which form natural wetlands in habitats favorable to their needs. Large, woody debris also forms natural dams that impound shallow wetlands.

Once formed, wetlands can also age over time, slowly filling in with external sources of materials such as sediments from upland erosion, as well as with detrital materials from wetland vegetation. Rates of deposition of these materials can be slow, such as in oligotrophic systems with small upstream catchment areas. Or they can be rapid, such as in nutrient-rich systems with large upstream areas marked by extensive erosion.

Wetlands in a natural setting, therefore, are constantly being formed and lost—depending on the balance of forces. So, too, wetland hydrology changes over time. Reducing the volume of storage within a wetland decreases the residence time, and can also reduce the depth and hydropattern by removing storage volume within the deep-water areas that would normally remain wet under drought conditions.

The dynamic nature of hydrology—especially when applied to wetlands—means that wetlands cannot be investigated apart from their regional environment. Hydrologic alteration upstream of the wetland affects wetland evolution.

Human Alteration of Wetland Hydrology

Humans have substantially increased hydrologic disturbances within watersheds. These changes generally cause increased sediment production and transport, as well as increases in nutrient concentrations and loads. Such increases naturally cause reductions in wetland storage volumes due to sediment trapping and nutrient uptake with subsequent deposition of organic sediments.

Surface water inflows to wetlands can be increased by routing stormwater runoff into them from urban, industrial, and agricultural areas. Inflows are also altered when hydraulic structures such as reservoirs, canals, levees, dikes, revetments, and jetties obstruct or alter natural hydrologic patterns. Many of these alterations resulted from efforts to drain wetlands.

Outflows from wetlands were increased by the construction of drainage ditches, channels, and canals, or the removal of natural barriers such as vegetation and by straightening streams. Other efforts to drain wetlands used ground water extraction techniques such as underground tile drains and pumping wells that led to lowered ground water levels. Lowering of water tables can affect wetlands by increasing subsurface drainage from the wetland to the point of ground water extraction. Ditches and tile drains increase the discharge of shallow ground water, thus lowering water tables in the vicinity of the drain. Water levels increase with distance away from the drains, reaching a maximum midway between the drains. Tile drainage systems increase the rate of shallow ground water flow, thus favoring drier conditions within the wetland. Tile drainage systems are more effective for removing water resulting from low-intensity, long-duration storms, and are less effective for draining water resulting from short-duration, high-intensity storms.

Drainage of fields for agriculture may reduce surface water inflows, lower water tables, and reduce the seasonal period of soil saturation. Ground water pumping in the vicinity of the wetland can lead to a reduction in shallow aquifer water levels while irrigation may increase water levels, resulting in either decreases or increases in wetland water levels, respectively. The effects of regional ground water pumping tend to manifest themselves as slow (and in some cases, rapid) declines in regional, confined aquifer levels. These declines are then transmitted by reductions in diffuse upward leakage or direct connections to wetlands, resulting in the lowering of water levels in wetlands.

The effect of the alteration of channels and canals can be two-fold—not only do the new excavations convey more water out of the wetland, the spoils (i.e., the materials removed from the excavated areas) are commonly piled near the excavations and may concentrate or otherwise alter the natural drainage through the wetland.

Beaver ponds form natural wetlands that once dotted the landscape. Beaver trapping and eradication efforts may therefore have reduced the formation of new wetlands. Also, reducing the availability of large, woody debris may reduce wetland formation. Harvesting of riparian vegetation—particularly the larger diameter trees—could result in poorer recruitment of large, woody debris. While wetlands clearly affect vegetation, fish, and wildlife, it is also true that these biological factors affect wetlands.

Obstructions (such as beaver dams, roads, channels, dams) to surface water exchanges alter the hydrology by requiring a higher stage to pass the same flow. Obstructions to inflows may deprive the wetland of natural flows. In some cases, obstructions may not substantially alter total wetland outflows; they may just alter the stage-discharge relationship, requiring a higher water level in order to pass an equivalent discharge. This may have both positive and negative effects on wetlands. Increased water levels can alter the natural storage ability (a negative effect), but may increase the residence time (a positive effect).

Some wetlands can have a large hydraulic effect by mitigating flood flows. Wetlands can retard floods by slowing the average water velocity, as well as the flood peak velocity. Rapid flood waves cause greater damage downstream because they have less time to dissipate their peaks. Slower floods have smaller peaks and lower velocities, resulting in less downstream damage. This is accomplished by the additional friction, or resistance to flow, that wetlands provide, along with the increased water storage capacity associated with their area.

Efforts to mitigate stormwater runoff have resulted in endeavors to design and construct artificial wetlands to mimic the beneficial hydrologic effects of natural wetlands. These endeavors focus on using hydrodynamic models to achieve specific management goals that require reductions in nutrients, sediments, and peak flows.

Manning's equation suggests that several factors affect the water velocity, including the flow roughness of flooded ground, the hydraulic radius (i.e., the effective water depth), and the water energy slope. Wetlands with substantial macrophytic vegetation can increase the hydraulic roughness, thus decreasing flow velocities. Also, shallow water bodies reduce the hydraulic radius, again decreasing flow velocities. Removing wetland vegetation thus decreases the hydraulic radius, causing increased water velocities.

It is also apparent that increased wetland loading rates, along with decreased retention times, substantially decreases the effectiveness of wetlands in storing water during flood periods and subsequently releasing it during dry periods. The resulting hydrologic performance of affected wetlands is fundamentally compromised.

Irrigation can increase water tables if the amount of irrigation exceeds plant needs. In many arid areas, surplus irrigation is required to remove dissolved salts from the root zone. Under these conditions, recharge from irrigation, precipitation, and/or surface flows may increase water levels to the point where surface inundation results, forming saline wetlands.

Some coastal wetlands are dominated by nearby freshwater inputs, especially in estuarine environments. In these cases, occasional, large storm water-driven inflows can cause rapid changes in the salinity, temperature, dissolved oxygen, and sediment concentration within the wetland. Alteration of the coastal morphology by dredging can adversely affect natural wetlands by increasing saltwater intrusion rates. Density-dependent stratification of estuarine waters may prevent salt-water presence in coastal areas. Construction of deep-water navigation channels may allow for salt-water to gain inland access, which can result in increased salinities.

Coastal ground water pumping affects coastal wetlands by reducing artesian pressures in underlying confined aquifers, which may then cause a reduction in point and diffuse upward leakage. Also, ground water pumping may cause coastal subsidence, resulting in the effective lowering of the ground surface relative to the sea level, causing the intrusion of saline water into coastal wetlands.

Pumping from shallow aquifers can also lower coastal zone water levels, causing local dewatering of coastal wetlands. Shallow disposal of septic wastes can alter local ground water quality by increasing organic and nutrient loading and decreasing dissolved oxygen concentrations. These changes can affect local wetlands if and when this ground water discharges into them.

Wetland Restoration

Efforts toward restoration of the hydrologic function of compromised wetlands are currently expanding. Additional efforts are being undertaken to create artificial wetlands that take advantage of the natural functions that wetlands provide. Regardless of whether impaired wetlands are being restored or new wetlands are being created, the intent is to recreate the hydrologic behavior that we find so important. The emphasis in these cases is the design and evaluation of alternative strategies for wetland restoration.

Water levels in wetlands can be controlled by manipulating the stage discharge relationship. Changing the elevation of an outflow structure, e.g., by raising the base of an outlet weir elevation, changes the wetland water levels and flooded areas. The base elevation, along with the rate of change in discharge with elevation, can be adjusted using outflow structures of different sizes and shapes, depending upon the desired outflow characteristics.

Other hydrologic alteration possibilities include closing of ditches and drains, thus reducing wetland outflows. Removing artificial obstructions such as roads and berms can also improve flow through the wetland by recreating natural hydrologic communication with neighboring water bodies.

These principles apply not only to freshwater wetlands but also to the restoration of tidal wetlands, which requires the recreation or simulation of natural hydraulic conditions. Weirs and plugs are devices used in tidal marshlands to maintain minimum water levels. Weirs are useful because the bottom elevation of the weir controls the minimum elevation on the upstream side but allows higher flows to pass unaffected. Ditch plugs provide the same control, but are more susceptible to destruction during high flows.

The term wetland hydrology generally refers to the inflow and outflow of water through a wetland and its interaction with other site factors. Land is characterized as having wetland hydrology when, under normal circumstances, the land surface is either inundated or the upper portion of the soil is saturated at a sufficient frequency and duration to create anaerobic conditions. The presence or absence of wetland hydrology may be determined through the on-site identification of established field indicators. While field indicators of wetland hydrology are at times difficult to identify, it is essential to determine that the area is periodically inundated or has saturated soils in order for the area to be characterized as meeting the wetland definition. For sites where there is a predominance of wetland plant species but there is no direct visible evidence that water is, or has been, at or above the soil surface, Part 303 directs the MDEQ to use the characteristics of soils to verify the presence or absence of wetland hydrology. If field indicators of hydrology are absent, such as in disturbed areas, evidence of hydrology may need to be established through the evaluation of recorded hydrologic data.

Site Factors that Influence Wetland Hydrology

Wetland conditions occur where topographic and hydrogeologic conditions are favorable and a sufficient, long-term source of water exists. Favorable topographic conditions refer generally to the presence of land-surface depressions in the drainage basin. These depressions may be located in upland areas, along hillsides where there may be a change in slope or geology, in floodplains of streams or rivers, or along the margins of lakes. Geologic conditions which may be favorable for wetland development include areas that have fine textured surficial soils with low hydraulic conductivity and sufficient thickness to store water. Also, the presence of impermeable bedrock near the land surface may favor the development of wetland hydrology.

The development of wetland conditions requires a persistent, long term source of water. Figure shows the different sources of water and possible outflows from a hypothetical wetland. The source of water may be precipitation (P) which falls directly on the wetland, surface water runoff during rainfall or snowmelt events within the catchment area surrounding the wetland (surface water inflow, or SWI), periodic flooding caused by elevated water levels in nearby surface water bodies (also SWI), groundwater inflow to the wetland (GWI), or a combination of any, or all, of these sources. Water may be lost from a wetland by evaporation from standing water or saturated soils (E), transpiration from plants (T), or surface water or groundwater outflow (SWD or GWD).

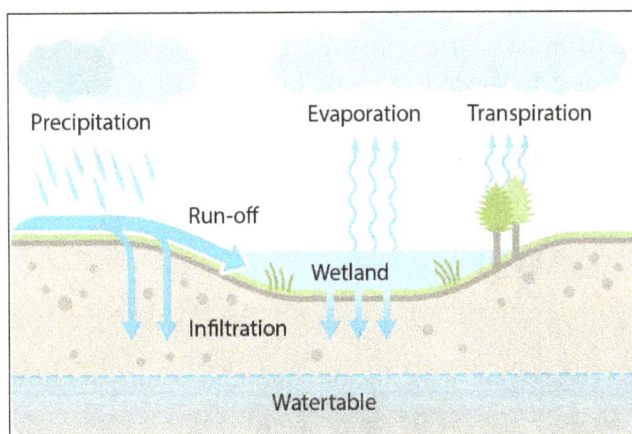

Wetland Hydrologic Cycle

The development of wetland conditions depends on a long-term balance between water inflow to the wetland and outflow from the wetland. During dry climatic periods, the rate of water inflow to the wetland (precipitation, groundwater inflow, and surface or near-surface inflow) may greatly diminish. In this instance, the amount of water lost through evapotranspiration may exceed the rate of all water inflow to the wetland. Water losses through evapotranspiration can result in extreme declines in the water table and a de-saturation of the wetland.

The relative importance of water inflow and water outflow, along with the topographic and geologic setting, determines the type and characteristics of the wetland that may form at a given location. A number of wetland classification systems have been developed that group wetlands based on topographic position in the landscape, water source, and hydrodynamics. Four commonly found wetland systems in Michigan are surface water depression wetlands, groundwater slope wetlands, groundwater depression wetland, and surface water slope wetlands. Wetlands that receive water primarily from precipitation have been classified as surface water depressional wetlands. Wetlands for which groundwater is the predominant source of water are classified as groundwater slope or groundwater depressional wetlands. Wetlands which are dependent upon surface water inflow are classified as either riverine or fringe wetlands along existing bodies of open water.

Figure shows a wetland that has formed in a topographic depression. The primary sources of water are precipitation and surface water runoff from the catchment area surrounding the wetland. Since the water level elevation in the wetland is greater than the elevation of the water table, water in the wetland moves toward the water table, and groundwater is not a source of water for the wetland. The outflow of water from this category of wetlands is evaporation from the water surface, transpiration from plants, and movement of water to the underlying or adjoining aquifer. The soils or geologic sediments which underlie the wetland may be predominantly clay. The relatively low hydraulic conductivity of the sediment restricts, but does not prohibit, the movement of water from the wetland to the underlying aquifer. This category of wetlands is referred to as surface water inflow or depressional wetlands. This category of wetlands may be found at any elevation, even in otherwise predominantly upland areas. These wetlands are more dependent on precipitation than other types of wetlands.

Hillslopes between upland and lowland areas are another topographic setting in which wetlands may form. Wetlands forming in these areas are referred to as groundwater slope wetlands. An example of this type of wetland is shown in figure. Groundwater which discharges along the hillslope as a seep or spring is the primary source of water to this wetland. Overland flow and precipitation may also contribute water to these wetlands. In this setting, sediments which have relatively low hydraulic conductivity such as clay or silt may underlie more permeable saturated sediments, forming a perched aquifer. Groundwater would flow laterally, along the clay or silt layer, toward the hillslope, where it discharges as a seep or spring. This is referred to as a groundwater seepage face. Groundwater slope wetlands may also occur where there are changes in the hillside slope and may not have perched groundwater conditions.

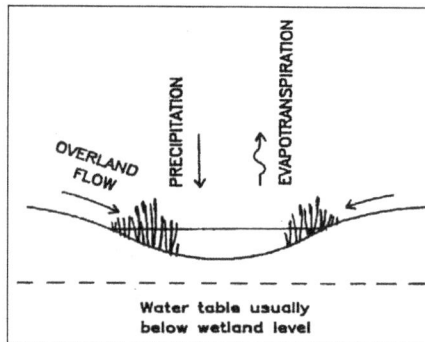

Surface water depression wetland

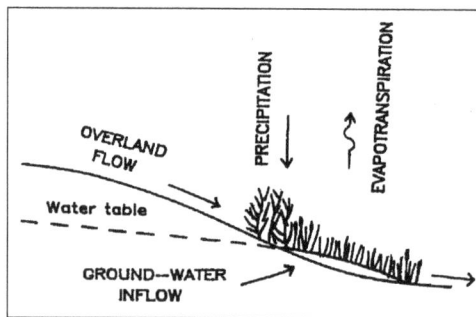

Groundwater slope wetland

Groundwater slope wetlands tend to have relatively constant inflow of water if the aquifer responsible for the water source is readily recharged or groundwater moves through the aquifer at a relatively high rate. In this case, the wetland would be relatively unaffected by seasonal demands by evapotranspiration. If a shallow perched aquifer provides water to the seep, the wetland soils may become dry during portions of the growing season because of evapotranspiration in the seepage area.

Groundwater slope wetlands generally have a surface water outlet. The size of these wetlands depends on the quantity of groundwater discharge and the slope of land surface down gradient of the seepage face or spring.

Figure shows a wetland formed in a topographic depression which may be in a lowland area. For this category of wetlands, the primary sources of water are groundwater discharge to the wetland, precipitation, and surface water runoff from the catchment area surrounding the wetland. Since the water table elevation is higher than the water level elevation in the wetland, groundwater moves from the adjoining and underlying aquifer toward the wetland. The outflow of water is from evaporation from the water surface and transpiration from plants. These wetlands may not have any surface water outlets. This category of wetlands is referred to as groundwater depression wetlands. While they can exist at any elevation, these wetlands are typically found in relatively low-lying areas.

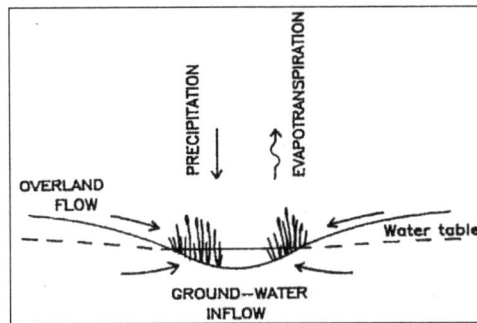

Groundwater depression wetland

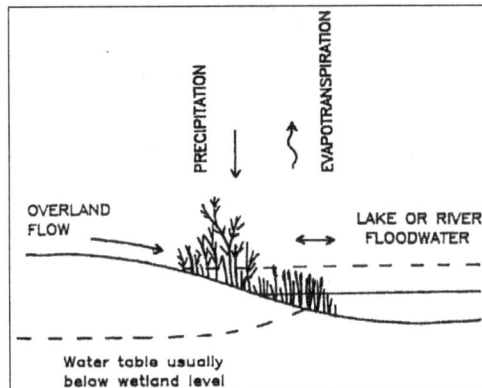

Surface water slope wetland

Another category of wetlands is referred to as surface water slope wetlands. Surface water slope wetlands receive water primarily from the flooding of lakes or rivers, and the water can readily drain back into lakes or rivers as the surface water stages decline. Within floodplains, the flooding occurs infrequently. However, lakeside wetlands may be flooded permanently. These areas near surface water bodies are generally areas of regional or local groundwater discharge. The discharging groundwater is an important, consistent source of water to these wetlands.

Riverine wetlands form as linear strips, generally paralleling river and stream channels. These wetlands are found at lower elevations in a floodplain and tend to be more frequently inundated and for a longer duration than areas at slightly higher elevations.

Fringe wetlands occur adjacent to lakes where water moves in and out of the wetland from the effects of wind, waves, and seiches. This is especially true for wetlands that have formed near the Great Lakes. Lakes that are too small to develop frequent seiches would not support fringe wetlands; such lakeside wetlands would fall into either the surface water or groundwater inflow category.

It is possible for the source of water to wetlands to change during wet and dry climatic cycles. As an example, the wetlands shown in figures might depict the same wetland, but under different climatic conditions. Figure, with the low water table and inflow by surface water and precipitation only, may represent relatively dry or drought conditions. The conditions shown in figure, with the high water table and surface water, precipitation, and groundwater inflow, may represent the same wetland during wet climatic periods.

Field indicators of Wetland Hydrology

The following field indicators of wetland hydrology are from the USACE manual. The indicators can be quickly assessed and provide support that inundation or soil saturation has occurred at a site. Although some indicators are not necessarily indicative of hydrologic events that only occur during the growing season, they do provide evidence that inundation and/or soil saturation has occurred at a site. The use of these field indicators requires on-site observations.

Primary indicators of Wetland Hydrology

- Visual observation of inundation - The most obvious and revealing hydrologic indicator may be simply observing the areal extent of inundation. However, because seasonal conditions and recent weather conditions can contribute to surface water being present on a non-wetland site, both should be considered when applying this indicator.

- Visual observation of soil saturation - Examination of this indicator requires digging a soil pit to a depth of 16 inches and observing the level at which water stands in the hole after sufficient time has been allowed for water to drain into the hole. The required time will vary depending on soil texture. In some cases, the upper level at which water is flowing into the pit can be observed by examining the wall of the hole. This level represents the depth to the water table. The depth to saturated soils will always be nearer the surface due to the capillary fringe.

- For soil saturation to impact vegetation, it must occur within a major portion of the root zone (usually within 12 inches of the surface) of the prevalent vegetation. The major portion of the root zone is that portion of the soil profile in which more than one half of the plant roots occur. In some heavy clay soils, water may not rapidly accumulate in the hole even when the soil is saturated. If water is observed at the bottom of the hole but has not filled to the 12-inch depth, examine the sides of the hole and determine the shallowest depth at which water is entering the hole. When applying this indicator, both the season of the year and preceding weather conditions must be considered.

- Watermarks - Watermarks are most common on woody vegetation. They occur as stains on bark or other fixed objects (e.g., bridge pillars, buildings, fences, etc.). When several watermarks are present, the highest reflects the maximum extent of recent inundation.

- Drift lines - This indicator is most likely to be found adjacent to streams or other sources of water flow in wetlands. Evidence consists of deposition of debris in a line on the surface or as debris entangled in above ground vegetation or other fixed objects. Debris usually consists of remnants of vegetation (branches, stems, and leaves), sediment, litter, and other waterborne materials deposited parallel to the direction of water flow. Drift lines provide an indication of the minimum portion of the area inundated during a flooding event; the maximum level of inundation is generally at a higher elevation than that indicated by a drift line.

- Sediment deposits - Plants and other vertical objects often have thin layers, coatings, or depositions of mineral or organic matter on them after inundation. This evidence may remain for a considerable period before it is removed by precipitation or subsequent inundation. Sediment deposition on vegetation and other objects provides an indication of the minimum inundation level. When sediments are primarily organic (e.g., fine organic material, algae), the detritus may become encrusted on or slightly above the soil surface after dewatering occurs.

- Drainage pattern within wetlands - This indicator, which occurs primarily in wetlands adjacent to streams, consists of surface evidence of drainage flow into or through an area. In some wetlands, this evidence may exist as a drainage pattern eroded into the soil, vegetative matter (debris) piled against the thick vegetation or woody stems oriented perpendicular to the direction of water flow, or the absence of leaf litter. Scouring is often evident around roots of persistent vegetation. Debris may be deposited in or along the drainage pattern.

  Drainage patterns also occur in upland areas after periods of considerable precipitation; therefore, topographic position must also be considered when applying this indicator.

Supplemental Indicators of Wetland Hydrology

In addition to the primary indicators of wetland hydrology, the MDEQ may also consider the following site conditions as supplemental indicators to support evidence of wetland hydrology:

- Oxidized rhizospheres (root channels) associated with living plant roots in the upper 12 inches of the soil Oxidized rhizospheres surrounding living roots are acceptable hydrology indicators on a case-by-case basis and may be useful in groundwater systems. Use caution that rhizospheres are not relicts of past hydrology. Rhizospheres should also be reasonably abundant and within the upper 12 inches of the soil profile. Oxidized rhizospheres must be supported by other indicators of hydrology, such as the FAC-neutral test if hydrology evidence is weak.

- Water-stained leaves - The presence of stained vegetation can be used as a secondary indicator of wetland hydrology. The physical appearance of leaves resulting from the continued presence of water and anaerobic processes will often darken leaf surfaces. Comparisons can be made to leaves occurring within obvious areas of upland vegetation.

- Local soil survey hydrology data for identified soils - In groundwater-driven systems, which lack surface indicators of wetland hydrology, it is acceptable to use local NRCS soil survey information to evaluate the hydrology parameter in conjunction with other information, such as the FAC-neutral test. Use caution in areas that may have been recently drained.

- FAC-neutral test - The FAC-neutral test results in a positive secondary indicator of hydrology when more of the dominant plant species have a wetland indicator category that is wetter than FAC. The FAC-neutral test considers FAC species (FAC-, FAC, or FAC+) as neutral and does not utilize them. Rather, the abundance of OBL, FACW+, FACW, and FACW- species are weighed against the abundance of UPL, FACU-, FACU, and FACU+ species (OBL + FACW species > FACU + UPL species) to determine whether the vegetation meets the FAC-neutral test.

- Bare soil areas - Wetlands that contain standing water for a relatively long duration may have areas of bare or essentially bare soil. Bare soil areas can be a result of surface flows carrying away ground litter or the presence of standing water within local depressions for a relatively long time with limited inputs of plant litter material.

- Morphological plant adaptations - Some plant species have recognizable physical characteristics, such as butressed trunks, that reflect their ability to occur and survive in wetland conditions.

Forest Hydrology

A forest may be defined as a biological community dominated by trees and other woody vegetation.

Forest Types

Forests can be classified according to a wide number of characteristics, with distinct forest types occuring within each broad category. However, by latitude, the three main types of forests are tropical, temperate, and boreal.

Tropical

Most tropical forests receive large amounts of rain annually (up to 100 inches), which is spread evenly throughout the year. However, there are some tropical forests that receive seasonal rainfall and experience both a wet and dry season. While tropical forests have many layers, most of the nutrients are held in the vegetation within the canopy; therefore, the soils are typically low in both mineral and nutrient content. Shallow roots allow for 'catching' any nutrients released by decaying leaves and ground litter.

Tropical forests are particularly important since they are unusually rich in biological diversity, especially insects and flowering plants. This incredible amount of biodiversity—accounting for 50 to 80 percent of the world's plant and animal species, with a potential for millions still undiscovered—is what defines these forests and makes them most unique. In just a few square kilometers, hundreds—even thousands—of tree and plant species can be found.

Deforestation is one of the greatest concerns in tropical areas, especially within rainforests which cover only a small area (approximately 7 percent) of the Earth's surface. Aside from their vast biodiversity, tropical forests provide homes to a large number of indigenous people. And, in looking beyond the typical forest offerings, tropical forests supply both local and global markets with a variety of ingredients for medicines; nearly half of all medicines used today are linked to discoveries within these forests.

Temperate

Temperate forests—common throughout North America, Eurasia, and Japan—are primarily deciduous, characterized by tall, broad-leafed, hardwood trees that shed brilliantly colored leaves each fall. These forests experience varied temperatures and 4 seasons, with winter often bringing below freezing temperatures and summer bringing higher heat and humidity. Rainfall also varies, averaging 30 to 60 inches annually, allowing for soils that are well developed and rich in organic matter. They also provide habitat for a wide variety of smaller mammal species, including squirrels, raccoons, deer, coyotes and black bear and many bird species, including warblers, woodpeckers, owls, and hawks.

Temperate forests are often most affected by human activity since they are located in or near the most inhabitable areas. The land in these areas has long been used for agriculture and grazing, although great expanses of forest regeneration and small areas of pristine forest exist. The hardwoods are valuable for making furniture and other commodities, and many remaining forests have been modified to accommodate recreation and tourism.

Boreal

Boreal forests (also known as taiga) are located just south of the tundra and stretch across large areas of North America and Eurasia. They are one of the world's largest biomes, encompassing about 11 percent of Earth's land area, but have very short growing seasons with little precipitation and represent relatively few tree species. The forest is dominated by coniferous trees, which have needle-shaped leaves with minimal surface area to prevent excessive water loss. These forests provide habitat for a few large mammal species, such as moose, wolves, caribou, and bears, and numerous smaller species, including rodents, rabbits, lynx, and mink.

Despite the remote locations and often inhospitable environment, boreal forests have long been a source of valuable resources. Fur trading began in the 1600s and continued well into this century. Boreal forests are also rich in metal ores—including iron—and coal, oil, and natural gas. Most importantly, the forest serves as a major source of industrial wood and wood fiber, including softwood timber and pulpwood. However, the low productivity rate in these forests leads to a slow rate of forest regeneration.

Forest hydrology combines aspects of two separate disciplines: hydrology and forestry. Hydrology is the science that studies the waters of Earth. Hydrology seeks to understand where water occurs; how water circulates; how and why water distribution changes over time; the chemical and physical properties of water; and the relation of water to living organisms.

Forestry is the science that seeks to understand the nature of forests and the interactions between the parts comprising a forest. Forest management started in the 1700s as a form of large-scale farming to improve yields of timber and fiber from forests. In the United States, watershed protection has been an integral part of forest management since its origins. The Organic Administrative Act of 1897 stated that forest reserves were to protect and enhance water supplies, reduce flooding, secure favorable conditions of water flow, protect the forest from fires, and provide a continuous supply of timber. The 1911 Weeks Act authorized the acquisition of federal lands in the East for the express purpose of protecting the watersheds of navigable waterways.

In recent decades, forestry has adopted more of an ecosystem management approach while still including timber production as an important goal. Although a forest is an ecosystem dominated by trees, a healthy forest includes other plants as well as soil, terrestrial and aquatic animals, and water—plus people who use the forest and its resources. Modern forest management therefore requires not only an understanding of forest science, soil science, and hydrology, but also principles of wildlife biology, land-use planning, and recreation planning.

Water and Forested Ecosystems

Ecologists consider water to be the defining part in an ecosystem, including the forest ecosystem. Water shapes the physical landscape through erosion and deposition. It also shapes the biological parts of the ecosystem by its presence or absence; its quantity and quality; and its occurrence and distribution. The water cycle plays a key role in ecosystem functions and processes.

Undulating forest surrounds meadows, lakes, and rivers of the Charlevoix Conservation Area, a coastal watershed in Quebec, Canada. This region, which illustrates a high-quality forested watershed, has been designated a biosphere.

Forests, in turn, are vital to the water cycle and to water quality. In essence, the forest acts like a giant sponge, filtering and recycling water. Approximately 80 percent of U.S. fresh-water resources are estimated to originate in forests, which cover one-third of the U.S. land area.

Tree leaves intercept water from rain, snow, and fog; the leaves also release water back to the atmosphere by evapotranspiration. Tree roots extract water from the soil while helping hold the soil in place. Forested land reduces the surface impact of falling rain through interception and delay of water reaching the surface. Forestland also decreases the amount and velocity of storm runoff over the land surface. This in turn increases the amount of water that soaks into the ground, a portion of which can ultimately recharge underlying aquifers. Conversely, water from hydraulically connected surficial aquifers may enter streams and wetlands helping to maintain their water levels during dry periods.

Forests and the Hydrologic Cycle

The surface water in a stream, lake, or wetland is most commonly precipitation that has run off the land or flowed through topsoils to subsequently enter the waterbody. If a surficial aquifer is present and hydraulically connected to a surface-water body, the aquifer can sustain surface flow by releasing water to it.

In general, a heavy rainfall causes a temporary and relatively rapid increase in streamflow due to surface runoff. This increased flow is followed by a relatively slow decline back to baseflow, which is the amount of streamflow derived largely or entirely from groundwater. During long dry spells, streams with a baseflow component will keep flowing, whereas streams relying totally on precipitation will cease flowing.

Generally speaking, a natural, expansive forest environment can enhance and sustain relationships in the water cycle because there are less human modifications to interfere with its components. A forested watershed helps moderate storm flows by increasing infiltration and reducing overland runoff. Further, a forest helps sustain streamflow by reducing evaporation (e.g., owing to slightly lower temperatures in shaded areas). Forests can help increase recharge to aquifers by allowing more precipitation to infiltrate the soil, as opposed to rapidly running off the land to a downslope area.

Riparian Areas

The riparian zone is broadly defined as the area between a body of water and the upland parts of the landscape that are rarely flooded except under the most extreme conditions. But the term also can refer more specifically to the immediate streamside area.

Riparian areas represent less than 10 percent of most forest ecosystems, yet these areas often are the most productive portions. Compared to upland regions, riparian areas have more water available; the vegetation is more robust; the soils are deeper; the timber often is of higher quality; and the waterbodies have more shade. The riparian zone also may include wetlands bordering streams and lakes. This combination of factors makes riparian areas among the most heavily used portions of a forest. Riparian and wetland areas provide abundant and reliable forage for wildlife, as well as transportation corridors. They also may receive heavy human use for recreation.

Riparian zones also are attractive destinations for logging and for livestock grazing; as a result, riparian areas in forests are sometimes heavily damaged, especially in the forests of the arid American Southwest. Fortunately, riparian areas respond well to good management practices.

Aquatic Biodiversity

Forest lands and waters are vitally important in maintaining biodiversity and providing habitat for fish and wildlife, including threatened or endangered aquatic species. In the United States, over one-third of national forest lands are critical for maintaining aquatic biodiversity and protection of listed species.

For aquatic species, watersheds provide the basic unit of any conservation strategy. Many watersheds also contain isolated habitats with unique characteristics producing a high potential for rare species. Some species occur only near a single spring or in a single stream within a given watershed. Lands set aside to protect these unique habitats also benefit the entire watershed and its ecosystem.

Forest Management and Watershed Quality

Wind, fire, insects, and disease are all part of properly functioning, healthy ecosystems in watersheds. For example, natural fires, although temporarily devastating, periodically restore the balance between vegetation types, and release nutrients from the vegetation and soil. In contrast, widespread clear-cut logging and excessive or improper road-building can degrade watersheds, as can land uses such as ski runs and housing projects. Many human activities can increase overland runoff, resulting in more erosion of the land surface and concurrently reducing the amount of water that soaks in the ground to potentially reach nearby streams or recharge underlying aquifers.

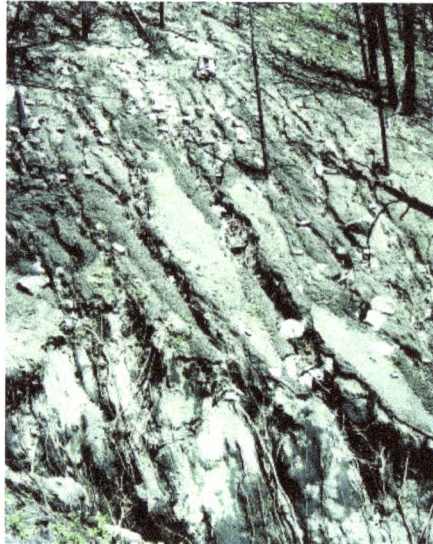

Rill and gully erosion caused by flowing water is evident on this steep
slope following a forest fire. The soil loss can be substantial,
and denuded slopes can be difficult to revegetate.

Moreover, fire prevention and suppression have created "imbalanced" forests with excessive amounts of undergrowth and dead vegetative matter that serve as fuels when fire does occur. Hence, these forests are at increased risk of high-intensity, destructive fires.

Watershed management and restoration may include controlled thinning, prescribed burning, and other management practices to restore the proper balance of timber, undergrowth, and grassy meadows in the watershed. Restoration also may include planting of appropriate native plants.

Effects of Roads

Improperly engineered roads in forests can increase erosion and significantly increase the risk of landslides. Both of the adverse effects are more severe when roads are numerous, and when they either cross or run parallel to streams. For example, heavy precipitation in Oregon and Washington during the mid-1990s resulted in many landslides, with a correlation between the slides and the frequency and density of roads. With respect to erosion and sedimentation, water runoff flowing along and across roads picks up sediment, which can then be deposited in nearby lakes and streams. This siltation can degrade or destroy habitat for aquatic organisms that require clear water and silt-free benthic (bottom) substrates.

Proper road engineering and following good practices (such as the U.S. Forest Service Guidelines for Best Management Practices) can reduce or eliminate the risk of erosion, landslides, and stream degradation due to excess siltation. Unfortunately, many roads in U.S. national forests were built before the practices were in place.

Effects of Fire

Destructive fires that remove large amounts of organic matter in a forest cause loss of nutrients from the soil as the detrital cover (i.e., dead and decaying materials on the forest floor) and upper soil layers are burned and eroded. Moreover, fires can adversely affect the quality of streams and lakes in the burned region as well as tributary watersheds downstream. Surface runoff that otherwise would have been slowed or absorbed by living and dead vegetative matter on the forest floor now runs unimpeded down bare (or nearly bare) slopes. The increased velocity carries more soil particles and loose vegetative matter downslope, along with any adsorbed nutrients.

References

- What-wetland, wetlands: epa.gov, Retrieved 9 April, 2019
- Wetlands-hydrology: epa.gov, Retrieved 14 March, 2019
- Water-wetlands, new-wetlands: forestandrange.org, Retrieved 5 August, 2019
- Forest-types, forests, land-use: enviroliteracy.org, Retrieved 25 January, 2019
- Forest-Hydrology: waterencyclopedia.com, Retrieved 30 July, 2019

Permissions

All chapters in this book are published with permission under the Creative Commons Attribution Share Alike License or equivalent. Every chapter published in this book has been scrutinized by our experts. Their significance has been extensively debated. The topics covered herein carry significant information for a comprehensive understanding. They may even be implemented as practical applications or may be referred to as a beginning point for further studies.

We would like to thank the editorial team for lending their expertise to make the book truly unique. They have played a crucial role in the development of this book. Without their invaluable contributions this book wouldn't have been possible. They have made vital efforts to compile up to date information on the varied aspects of this subject to make this book a valuable addition to the collection of many professionals and students.

This book was conceptualized with the vision of imparting up-to-date and integrated information in this field. To ensure the same, a matchless editorial board was set up. Every individual on the board went through rigorous rounds of assessment to prove their worth. After which they invested a large part of their time researching and compiling the most relevant data for our readers.

The editorial board has been involved in producing this book since its inception. They have spent rigorous hours researching and exploring the diverse topics which have resulted in the successful publishing of this book. They have passed on their knowledge of decades through this book. To expedite this challenging task, the publisher supported the team at every step. A small team of assistant editors was also appointed to further simplify the editing procedure and attain best results for the readers.

Apart from the editorial board, the designing team has also invested a significant amount of their time in understanding the subject and creating the most relevant covers. They scrutinized every image to scout for the most suitable representation of the subject and create an appropriate cover for the book.

The publishing team has been an ardent support to the editorial, designing and production team. Their endless efforts to recruit the best for this project, has resulted in the accomplishment of this book. They are a veteran in the field of academics and their pool of knowledge is as vast as their experience in printing. Their expertise and guidance has proved useful at every step. Their uncompromising quality standards have made this book an exceptional effort. Their encouragement from time to time has been an inspiration for everyone.

The publisher and the editorial board hope that this book will prove to be a valuable piece of knowledge for students, practitioners and scholars across the globe.

Index

www.ingramcontent.com/pod-product-compliance
Lightning Source LLC
Chambersburg PA
CBHW082022190326

41458CB00010B/3247